Christina Wolschner

Methionine Oxidation in Human Prion Protein

Christina Wolschner

Methionine Oxidation in Human Prion Protein

Design of Anti- and Pro-Aggregation Variants

Südwestdeutscher Verlag für Hochschulschriften

Impressum / Imprint

Bibliografische Information der Deutschen Nationalbibliothek: Die Deutsche Nationalbibliothek verzeichnet diese Publikation in der Deutschen Nationalbibliografie; detaillierte bibliografische Daten sind im Internet über http://dnb.d-nb.de abrufbar.

Alle in diesem Buch genannten Marken und Produktnamen unterliegen warenzeichen-, marken- oder patentrechtlichem Schutz bzw. sind Warenzeichen oder eingetragene Warenzeichen der jeweiligen Inhaber. Die Wiedergabe von Marken, Produktnamen, Gebrauchsnamen, Handelsnamen, Warenbezeichnungen u.s.w. in diesem Werk berechtigt auch ohne besondere Kennzeichnung nicht zu der Annahme, dass solche Namen im Sinne der Warenzeichen- und Markenschutzgesetzgebung als frei zu betrachten wären und daher von jedermann benutzt werden dürften.

Bibliographic information published by the Deutsche Nationalbibliothek: The Deutsche Nationalbibliothek lists this publication in the Deutsche Nationalbibliografie; detailed bibliographic data are available in the Internet at http://dnb.d-nb.de.

Any brand names and product names mentioned in this book are subject to trademark, brand or patent protection and are trademarks or registered trademarks of their respective holders. The use of brand names, product names, common names, trade names, product descriptions etc. even without a particular marking in this work is in no way to be construed to mean that such names may be regarded as unrestricted in respect of trademark and brand protection legislation and could thus be used by anyone.

Verlag / Publisher:
Südwestdeutscher Verlag für Hochschulschriften
ist ein Imprint der / is a trademark of
OmniScriptum GmbH & Co. KG
Heinrich-Böcking-Str. 6-8, 66121 Saarbrücken, Deutschland / Germany
Email: info@svh-verlag.de

Herstellung: siehe letzte Seite /
Printed at: see last page
ISBN: 978-3-8381-0940-4

Zugl. / Approved by: Munich, Technical University Munich, Thesis, 2009

Copyright © 2009 OmniScriptum GmbH & Co. KG
Alle Rechte vorbehalten. / All rights reserved. Saarbrücken 2009

Table of contents

I   **Nomenclature and Definitions** I
    I.I   Nomenclature I
    I.II   Definitions V

1   **Summary** 1
2   **Zusammenfassung** 2
3   **Introduction** 4
    3.1   DNA and the genetic information - historical background 4
    3.2   Flow of the genetic information and the genetic code 4
    3.3   The degeneracy of the genetic code 6
    3.4   Protein synthesis 7
        3.4.1   *Natural co-translation of selenocysteine and pyrrolysine* 8
    3.5   Engineering the genetic code 9
        3.5.1   *Comparison of different approaches* 12
    3.6   Methionine - a unique amino acid in the genetic code 14
        3.6.1   *Properties of methionine* 14
        3.6.2   *Methionine in the protein synthesis* 15
        3.6.3   *Methionine metabolism* 16
        3.6.4   *Methionine oxidation* 17
    3.7   Proteins and their conformational properties 19
        3.7.1   *Protein folding* 19
        3.7.2   *Protein misfolding and disease* 20
        3.7.3   *Oxidative stress and protein damage* 21
    3.8   Prion disease 22

4   **The Goal: Chemical model for prion protein conversion** 26
5   **Material** 27
    5.1   Equipment 27
    5.2   Chemicals 28
    5.3   Buffers and solutions 29

5.4	Polyacrylamide gel electrophoresis (PAGE)	30
5.5	Media	31
5.6	Enzymes	32
5.7	Protein molecular weight marker	32
5.8	Plasmids	32
5.8.1	pET17b-hPrP(23-231)WT81	32
5.8.2	pRIL	32
5.9	Antibiotics	33
5.10	Bacterial strains	33
5.11	Fluorescent dyes	33
5.12	Antibodies	33
5.13	Software	33
6	**Methods**	**35**
6.1	Microbiological methods	35
6.1.1	Production of electrocompetent cells	35
6.1.2	Transformation	35
6.1.3	Limitation test	35
6.1.4	Expression test	36
6.1.5	Protein fermentation and expression of full length $rhPrP^C$	36
6.1.6	Protein purification of full length $rhPrP^C$	36
6.1.7	Protein purity and concentration determination	37
6.2	Spectroscopy and spectrometry	38
6.2.1	Mass spectrometry (Orbitrap ESI-MS/MS)	38
6.2.1.1	In-solution Digestion.	38
6.2.1.2	NanoLC-MS/MS	38
6.2.1.3	Peptide Identification via MASCOT Search	39
6.2.2	Electrospray mass spectrometry (ESI-MS)	39
6.2.3	Circular dichroism (CD) spectroscopy and melting curves	39
6.2.4	Quantification of individual Met residue oxidation in $rhPrP^C$	40
6.3	Peptide synthesis	40
6.4	FCS/SIFT measurements and analysis	41
6.4.1	Fluorescent labeling of $rhPrP^C$	42
6.4.2	Quality control of labeled $rhPrP^C$	43

  6.4.3 *Western blot* 45
  6.4.4 *Aggregation assay* 45
  6.4.5 *Aggregation assay in different periodate concentrations* 46

7 Results 48

 7.1 Expression and purification of the rhPrP^C model protein 48
 7.2 Sodium periodate induced aggregation of Met-rhPrP^C 50
 7.3 Attempts to map Met oxidation by Orbitrap ESI-MS/MS 52
 7.4 Mapping Met oxidation by Bruckner Daltonics ESI-MS 55
  7.4.1 *Met-rhPrP^C oxidation using 5 equiv. NaIO$_4$* 55
  7.4.2 *Met-rhPrP^C oxidation using 25 equiv. NaIO$_4$ (soluble fraction)* .. 58
  7.4.3 *Met-rhPrP^C oxidized using 25 equiv. NaIO$_4$ (pellet fraction)* 60
  7.4.4 *Overall picture of NaIO$_4$ induced Met oxidation* 62
 7.5 Attempts for Met(O) and Met(O$_2$) incorporation into rhPrP^C 62
 7.6 Expression and isolation of Nle-rhPrP^C and Mox-rhPrP^C 64
 7.7 CD spectroscopy of Nle-rhPrP^C and Mox-rhPrP^C 68
 7.8 Monitoring secondary structure change upon heating 69
 7.9 Melting curves of Met-rhPrP^C and its variants 71
 7.10 Pro- and anti-aggregation prion protein variants 72
 7.11 Far-UV CD spectroscopy of the designed Nle and Mox peptide 74

8 Discussion 76

 8.1 Met oxidation as a possible origin of prion protein structural conversion... 76
 8.2 Recombinant hPrP^C as model for structural conversion 76
 8.3 Protein damage caused by Met oxidation 77
 8.4 Oxidation of Met residues in rhPrP^C and structural conversion 78
  8.4.1 *Met residues in the prion protein* 78
  8.4.2 *Structural consequences of Met oxidation in the prion protein* .. 79
  8.4.3 *Prion protein aggregation upon periodate oxidation* 79
 8.5 Chemical model for α → β conversion in rhPrP^C 82
 8.6 Proof of principle of the newly developed chemical model 84
 8.7 The model peptide.. 85

9 Conclusions 87

10	References	89
11	Figure List	101
12	Table List	104
13	Appendix	105
	13.1 Mapping NaIO$_4$ induced Met oxidation by mass spectrometry-5 equivalents NaIO$_4$	105
	13.2 Mapping NaIO$_4$ induced Met oxidation by mass spectrometry-25 equivalents NaIO$_4$- soluble fraction	112
	13.3 Mapping NaIO$_4$ induced Met oxidation by mass spectrometry-25 equivalents NaIO$_4$- pellet fraction	116

Nomenclature and Definitions

I Nomenclature and Definitions

I.I Nomenclature

Abbreviations used throughout the thesis are according to the recommendations of the IUPAC-IUB Commission of biochemical nomenclature and of the ACS Style Guide.

Furthermore, the following abbreviations were used:

A	Adenine
AARS	Aminoacyl-tRNA-synthetase
APS	Ammonium persulfate
Arg	Arginine
BHMT	Betaine:homocysteine methyltransferase
C	Cytosine
CD	Circular dichroism
CJD	Creuzfeldt-Jakob disease
DMG	*N,N*-dimethylglycine
E. coli	*Escherichia coli*
ε_M	Molar extinction coefficient
ER	Endoplasmatic reticulum
ESI-TOF	Electrospray ionization-time of flight
FCS	Fluorescence cross-correlation spectroscopy analysis
FFI	Fatal Familia Insomnia
fMet	Formylmethionine
G	Guanine
GPI	Glycosylphosphatidylinositol

Nomenclature and Definitions

GSS	Gerstmann-Sträussler-Scheinker syndrome
Hcy	Homocysteine
His	Histidine
Ile	Isoleucine
IPTG	Isopropyl-β-D-thiogalactopyranoside
λ_{Ex}	Excitation wavelength
LC	Liquid chromatography
MAT	Methionine adenosyltransferase
Met	Methionine
Mox	Methoxinine
Met(O)	Methionine sulfoxide
Met(O$_2$)	Methionine sulfone
MetRS	Methionyl-tRNA-synthetase
mRNA	Messenger RNA
MS	Met synthase
MS	Mass spectrometry
MsrA	Methionine sulfoxide reductase A
MsrB	Methionine sulfoxide reductase B
N	Number of particles
NADP	Nicotinamide adenine dinucleotide (oxidized)
NADPH	Nicotinamide adenine dinucleotide (reduced)
NaIO$_4$	Sodium periodate
nvCJD	New variant CJD
OD$_{600}$	Optical density at λ = 600 nm
PAGE	Polyacrylamide gel electrophoresis

Nomenclature and Definitions

Phe	Phenylalanine
PMDs	protein misfolded diseases
Pro	Proline
PrP^C	Cellular prion protein
PrP^{Sc}	Scrapie prion protein
Q-TOF	Quadrupole-time of flight
$rhPrP^C$	Recombinant human cellular prion protein
RNA	Ribonucleic acid
ROS	Reactive oxygen species
RP-HPLC	Reverse phase-high performance liquid chromatography
rRNA	Ribosomal RNA
rpm	Revolutions per minute
SAM	S-adenosylmethionine
SAH	S-adenosylhomocysteine
SAHH	S-adenosylhomocysteine hydrolase
SDS	Sodium dodecyl sulfate
SIFT	Scanning for intensely fluorescent targets
Ser	Serine
SPI	Supplementation incorporation method
T	Thymine
Tau	Taurine
TEMED	*N,N,N',N'*-tetramethylethylenediamine
THF	Tetrahydrofolat
tRNA	Transfer RNA
Trp	Tryptophan

Nomenclature and Definitions

TSE	Transmissible spongiform encephalopathies
Tyr	Tyrosine
U	Uracil
UV_{280}	Ultraviolet light at λ = 280 nm

Nomenclature and Definitions

I.II Definitions

Canonical amino acids are named according to the standard three letter code, e.g. methionine (Met).

Non-canonical amino acids are generally denoted by the standard three letter code e.g. norleucine (Nle) or methoxinine (Mox).

Mutant denotes a protein, in which the wild-type sequence is changed by site-directed mutagenesis in the frame of the 20 canonical amino acids.

Analog/Variant denotes a protein, in which one or more canonical amino acids from a wild type or mutant sequence are replaced by non-canonical ones. The respective incorporated non-canonical functional group is used as prefix of the protein name, e.g. recombinant human prion protein ($rhPrP^C$) with incorporated norleucine (Nle) or methoxinine (Mox), is called Nle-$rhPrP^C$ and Mox-$rhPrP^C$, respectively.

Equivalent denotes mol oxidant per mol protein.

1 Summary

The aim of this thesis was to examine the pathological relevance of the oxidation state of methionine (Met) side chains, in neurodegenerative disorders such as sporadic prion disease. Oxidative stress leading to Met oxidation is assumed to promote neurodegenerative disorders, as well as various aging processes by mediating protein misfolding and aggregation. The α-helix → β-sheet structural conversion of the cellular prion protein (PrP^C) into its misfolded and aggregated 'scrapie' (PrP^{Sc}) isoform is characteristic for prion disease. However, the exact pathogenic mechanism is still unknown.

First for better understanding the oxidative event, the effect and consequences of Met oxidation in the recombinant human cellular prion protein (Met-rhPrP^{C}23-231) were studied. Using sodium periodate induced Met oxidation of Met-rhPrP^{C}23-231, not only the extent of oxidation but also the identity of crucial residues for prion protein transition was determined. Furthermore, a concomitant increase in Met-rhPrP^{C}23-231 aggregation tendency with increasing periodate concentrations was observed in fluorescence cross-correlation spectroscopy.

Second the incorporation of Met analogs – mimicking the reduced and oxidized state – in rhPrP^{C}23-231 was performed. Due to the natural intracellular reduction of the Met residues, it was not possible to quantitatively introduce methionine sulfoxide into rhPrP^{C}23-231 and alternatives were required. Thus, the more hydrophobic norleucine (Nle) was used as non-oxidizable analog for Met and the highly hydrophilic methoxinine (Mox) as oxidized Met analog. Expectedly, the Nle-rhPrP^C variant is an α-helix rich protein like Met-rhPrP^C but resistant to oxidation and lacks the *in vitro* aggregation tendency of the parent protein. In contrast, the Mox-rhPrP^C variant is a β-sheet rich protein exhibiting strong pro-aggregation behavior. Both Nle/Mox-variants are not sensitive to periodate induced *in vitro* aggregation.

These results strongly indicate a correlation of the α → β secondary structure conversion in rhPrP^{C}23-231, with the oxidative state of the Met residues. In the future, this approach will certainly be useful for studying diseases, which arise from protein misfolding due to oxidative stress.

2 Zusammenfassung

Ziel dieser Arbeit war es, die pathologische Bedeutung des Oxidationszustands der Methionin-Seitenketten (Met) in neurodegenerativen Krankheiten, wie der sporadischen Prion-Erkrankung, zu untersuchen. Es wird vermutet, dass die durch oxidativen Stress verursachte Met-Oxidation zu Proteinfehlfaltung und Aggregation führt und dieses die Entwicklung von neurodegenerativen Erkrankungen sowie Alterungsprozessen fördert. Das charakteristische Merkmal der Prion-Erkrankung ist die α-Helix → β-Faltblatt Umwandlung vom zellulären Prion Protein (PrP^C), in die fehlgefaltete und zu Aggregation neigende 'Scrapie' (PrP^{Sc}) Isoform. Die genauen molekularen Hintergründe sind allerdings noch nicht geklärt.

Um die Auswirkungen der Oxidation besser verstehen zu können, wurden zunächst die Folgen der Met-Oxidation mittels Natriumperiodat im rekombinanten humanen zellulären Prion Protein (Met-rhPrP^C23-231) untersucht. Auf diese Weise war es nicht nur möglich das Ausmaß der Oxidation zu ermitteln, sondern auch die einzelnen oxidierten Met-Reste, die für die Prion-Umformung verantwortlich sein könnten, zu identifizieren. Mit Hilfe der Fluoreszenz-Kreuzkorrelations-Spektroskopie zeigte sich außerdem, dass steigende Periodatkonzentrationen mit einem erhöhten Aggregationsverhalten von Met-rhPrP^C23-231 einhergehen. Um das Protein im reduzierten und im oxidierten Zustand simulieren zu können, wurden entsprechende Met-Analoga in rhPrP^C23-231 eingebaut. Aufgrund der natürlichen intrazellulären Reduktion der Met-Reste war es nicht möglich, durch den Einbau von Methioninsulfoxid einen quantitativen und stabilen Oxidationszustand in rhPrP^C23-231 zu erzielen. Als Alternativen wurden Norleucin (Nle) als nicht oxidierbares Met-Analog und Methoxinine (Mox) als oxidiertes Analog verwendet. Wie erwartet hat die Nle-rhPrP^C-Variante einen hohen α-helikalen Anteil, wie auch Met-rhPrP^C, ist aber gegenüber Oxidation stabil und aggregiert kaum. Im Gegensatz dazu hat die Mox-rhPrP^C-Variante einen hohen β-Faltblattanteil und neigt zu Aggregation. Periodat hat *in vitro* keine Auswirkungen auf das Aggregationsverhalten der Prion-Varianten mit Nle oder Mox.

Zusammenfassung

Diese Ergebnisse deuten stark auf einen Zusammenhang zwischen der α → β-Umwandlung der Sekundärstruktur und dem Oxidationszustand der Methionin-Reste im rhPrP^C hin. In Zukunft wird diese Methode sicherlich bei der Untersuchung von Krankheiten, die mit Proteinfehlfaltungen durch oxidativen Stress einhergehen, äußerst nützlich sein.

3 Introduction

3.1 DNA and the genetic information - historical background

One of the fundamental similarities of all living beings is that they share a common chemistry. This chemistry is implemented by the catalytic activities of proteins that are encoded in the desoxyribonucleic acid (DNA). For a long time, people believed that proteins are the carrier of the genetic information, although F. Miescher already isolated DNA out of human cells in 1871 (1). It took until 1944 when O. T. Avery and his coworkers experimentally identified DNA as the carrier of the genetic information (2). Soon after, Levine and Todd discovered that the DNA molecule was built up of four bases (adenine (A), thymidine (T), guanine (G) and cytosine (C)), which are held together by a sugar-phosphate backbone. The next step was to clarify the question, how these four bases are combined, to store the genetic information. The most important experiment regarding this question was the discovery of the *Chargaff rules*, named after the Austrian chemist E. Chargaff. He discovered that the amount of A or G equals the amount of T or C, respectively. Using the information provided by E. Chargaff and the crystallographic information produced by R. Franklin, M. Wilkins together with J. Watson and F. Crick developed a model for the helical structure of DNA. The DNA molecule is a twisted ladder, where two long and twisted sugar-phosphate backbones form the outside of the ladder and the rungs are formed by the base pairs, A-T and G-C, which are weakly joined in the center by hydrogen bonds (3-7). This knowledge has enabled modern biology to make great leaps in understanding the human genome and the importance of DNA for life and health.

3.2 Flow of the genetic information and the genetic code

The concept known as the *'central dogma of molecular biology'* was originally proposed by F. Crick in 1957. This was the first hypothesis capable to predict a transfer direction of the genetic information in living organisms. The basic statement is that the flow of genetic information is unidirectional. This means that the transfer of information from nucleic acid to nucleic acid or from nucleic acid to protein is

Introduction

possible, but transfer from protein to protein or protein to nucleic acid is impossible (8). In this scheme, the genetic information stored in the DNA is first transcribed into RNA (where T is replaced by uracil (U)), by the cellular transcription machinery. Different forms of RNAs are known, and all of them exhibit different functions, e.g. messenger RNA (mRNA) is the template for protein synthesis, transfer RNA (tRNA) transports activated amino acids to the ribosome, ribosomal RNA (rRNA) forms the main component of the ribosomes, and small nuclear RNA (snRNA) participates in the splicing process on RNA exons. After transcription, the mRNA is translated into their corresponding amino acid sequence at the ribosome (9).

The crucial experiment to decode the genetic code was done in 1961 by M. Nirenberg and his colleague H. Matthaei (10), when they used a cell-free system to translate a poly-uracil RNA sequence and discovered that the polypeptide they had synthesized consisted of only the amino acid phenylalanine. Five years later the genetic code was completely deciphered. After the elucidation of the triplet code it became clear, how the information of linear amino acid sequences was preserved in co-linear nucleic acid sequences. This means that on the mRNA three nucleobases (coding triplets) encode for one amino acid in the related protein. The four bases A, U, G and C can be combined into $4 \times 4 \times 4 = 64$ distinct triplets. Three triplet combinations (UAA (ochre), UGA (opal) and UAG (amber)) are known to code for stop codons, which are signals for chain translation termination. The remaining 61 codons encode for a pool of 20 canonical amino acids that are invariant in all known organisms. Therefore, most but not all amino acids are encoded by more than one codon (Fig. 1).

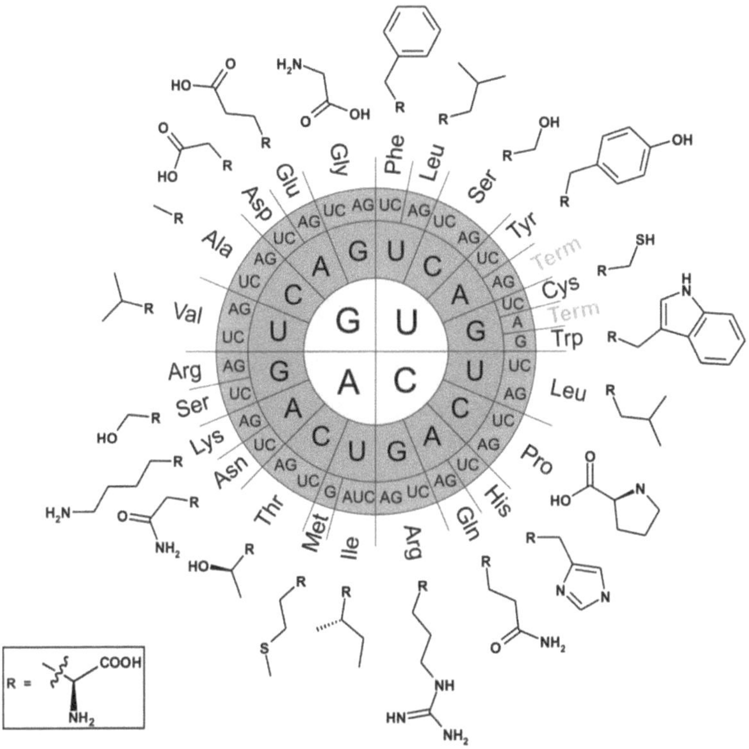

Introduction

Fig. 1 The standard genetic code (RNA format). 64 triplets are encoding for 20 different amino acids and three stop codons (denoted as Term; UAA, UGA, and UAG). Ala: Alanine; Arg: Arginine; Asn: Asparagine; Asp: Aspartic acid; Cys: Cysteine; Gln: Glutamine; Glu: Glutamic acid; Gly: Glycine; His: Histidine; Ile: Isoleucine; Leu: Leucine; Lys: Lysine; Met: Methionine; Phe: Phenylalanine; Pro: Proline; Ser: Serine; Thr: Threonine; Trp: Tryptophan; Tyr: Tyrosine; Val: Valine.

3.3 The degeneracy of the genetic code

The reason why the genetic code is degenerated or redundant in its structure is that 18 out of the 20 amino acids are encoded by two or more codons. Only two amino acids, methionine (Met) and tryptophan (Trp) are related to just one codon. The codon for Met (AUG) also acts as an initiation (start) codon. Codons that specify the same amino acid are called synonyms and they usually differ only in the third base. During protein synthesis, each codon is recognized by a triplet of bases

Introduction

(anticodon), in a specific tRNA molecule. In 1966 F. Crick formulated the 'wobble hypothesis', which postulates less strict specificity in the pairing of the 5' base of the tRNA anticodon with the 3' base of the codon. This allows alternative hydrogen bonding with the third base of the codon of the mRNA. Therefore, a point mutation in the DNA or mRNA that changes just the 3' nucleotide of a codon has often no effect on the amino acid sequence of the encoded polypeptide (11).

3.4 Protein synthesis

Normally, proteins consist only of 100 to 1000 amino acids and, hence, it is essential to keep translation as precise as possible. Misincorporation of amino acids into the polypetide chain during translation occurs only in one out of 10^4 cases (12). The adapters, which are responsible for the correct connection between nucleic acid and amino acid, are the tRNAs. Their secondary structure is usually represented by a cloverleaf. It consists of three stem loops, one of which bears the anticodon. The amino acid is attached onto the acceptor part on the opposite of the anticodon stem. A cognate enzyme called aminoacyl-tRNA-synthetase (AARS) catalyzes the attachment of an amino acid to a particular tRNA. For each canonical amino acid are usually a few tRNAs, but only one AARS. This enzyme catalyzes the aminoaclyation reaction, which takes place in two steps. In the first step the enzyme recognizes its 'own' amino acid and binds ATP (cofactor Mg^{2+}), which further forms aminoacyl-adenylate (aminoacyl-AMP) by concomitantly releasing pyrophosphate (activation step). In the second step, the enzyme binds the cognate tRNA whereby the aminoacyl group of aminoacyl-AMP is transferred to the 3' end of the tRNA to form the charged-tRNA or aminoacyl-tRNA (transfer step). The synthesis of aminoacyl-tRNAs is the crucial step in protein biosynthesis, because each amino acid must be covalently linked to a tRNA adapter in order to take part in protein synthesis at the ribosome. Additionally this covalent bond is a high energy bond, which enables the amino acid to react with the end of the growing polypeptide chain to form a new peptide bond. During translation the ribosome reads the nucleotide sequence in the 5' to 3' direction, while synthesizing the corresponding protein from amino acids in an N-terminal to C-terminal direction (9, 11, 12).

Introduction

Protein synthesis at ribosome takes place in three steps: initiation, elongation and termination. During initiation the mRNA-ribosome complex is formed and the first codon (AUG) binds the aminoacyl-tRNA$_{CAU}$ (which is N-formylmethionine (fMet)-tRNA$_{CAU}$ in prokaryotes and Met-tRNA$_{CAU}$ in eukaryotes). In the elongation phase, the other codons are sequentially read and the polypeptide grows by amino acid addition to its C-terminal end. In the last phase, known as termination, the ribosome reaches a stop codon, which does not have a corresponding aminoacyl-tRNA, but related release factor that terminates chain elongation. Thereby protein synthesis ceases and the completed polypeptide is released from the ribosome (9). Approximately 10-20 polymerization reactions are performed per second by the ribosome, therefore polypeptide biosynthesis is five times slower than RNA synthesis (13).

The ribosome contains three binding sites for the tRNA. These are called A- (aminoacyl-), P- (peptidyl-) and E- (exit-) site. While a tRNA is connected with the nascent peptide chain at the P-site, an aminoacyl-tRNA binds to the A-site. The amino group of the aminoacyl-tRNA nucleophil attacks the carbonyl-group of the peptidyl-tRNA in such a way that a new peptide bond is formed. After this process, the tRNA and the mRNA have to be moved on so that a new cycle can start. The deacetylated tRNA moves to the E-site where it leaves the ribosome, and the peptidyl-tRNA moves from the A to the P-site (9).

In all living organism protein synthesis is performed by the ribosomes and consumes a lot of energy. Moreover, a minimum of 35-45% of the genome is assigned to the protein synthesis apparatus. In humans, protein synthesis burns up approximately 5% of the caloric intake and in *E.coli* it even uses 30-50% of the generated energy (14). In living cells the cellular investment in protein synthesis can be correlated to the amount of RNA, since all RNA types account for about 21% of the whole biomass in an average *E.coli* population (14).

3.4.1 Natural co-translation of selenocysteine and pyrrolysine

Though all living organisms translate the same 20 canonical amino acids into protein sequences according to the rules of the genetic code, some deviation in its interpretation can be found. In mitochondria, chloroplasts and in some ciliates, the assignment of codons to amino acids is changed, e.g. UAA and UAG code for glutamine instead of translation termination (11). Moreover, two additional amino acids, beside the 20 canonical amino acids, named selenocysteine (15) (in archea, bacteria and mammals) and pyrrolysine (16) (in archea), can be incorporated into the nascent polypeptide chain. However, their incorporation is sequence context dependent and requires both, specific translation factors and specific secondary structure elements in the mRNA.

3.5 Engineering the genetic code

By the end of the 19th century most of the amino acids were identified, which are known to be the main buildings blocks of all proteins. 50 years later, there were considerable interests to study cellular metabolism and growth by using non-canonical amino acids. The most important and far-reaching finding was the quantitative replacement of Met by selenomethionine (SeMet) in proteins, reported by Cowie and Cohen (17). For this incorporation experiment, the use of a Met auxotrophic *E.coli* strain and defined fermentation conditions were mandatory. Amino acid auxotrophic bacteria strains will proliferate only if the medium is supplemented with the appropriate amino acid (e.g. the canonical amino acid Met). Therefore, growth of the culture is fully dependent on the external Met supply, and in a defined synthetic medium, it is possible to replace Met by SeMet. This landmark experiment was the first and is till date the only experiment with a proteome wide canonical amino acid → noncanonical analog substitution, none the less the cell capacity to divide and grow was retained. In 1990 W. Hendrickson rediscovered the SeMet incorporation into proteins, using it as an important tool for structural biology, especially for X-ray and NMR analyses of biological macromolecules (18).

Cowie and Cohen reported that the incorporation of SeMet into *E.coli* proteins was almost quantitative and the cellular viability was not significantly harmed. However, this is rather an exception. Already in early experiments, it became clear that non-canonical amino acids are mainly bacteriostatic or even bactericidal. For example, the Met analog Nle leads to only 38% replacement of Met in proteins accompanied by greatly impaired viability of *E.coli* ML304d (17, 19). Therefore, to circumvent the

Introduction

problem of analog toxicity and to achieve full substitution of single target proteins, the translation capacity should be resolved from metabolic toxicity (20). For this approach, auxotrophic host cells are starved on a specific canonical amino acid and protein synthesis is induced with concomitant addition of the analog to the culture medium. This approach together with routine recombinant DNA techniques (i.e. heterogonous gene expression in a highly efficient and controllable manner) allows the production of a single substituted target protein, without the need for large scale changes, in the host cell proteome. Consequently, the basic prerequisites for successful non-canonical amino acid incorporation into a defined target protein are, first bacterial expression of host cells with a genetically stable auxotrophic marker for the amino acid that should be exchanged; second a tightly controlled but highly inducible expression vector for the gene of the model protein; and finally a calibrated fermentation, where the cell mass is grown on a limiting amount of canonical amino acid that is exhausted as soon as the cells enter mid-log phase. After this amino acid is depleted from the media, the non-canonical amino acid is added and target gene synthesis is induced. Thereupon, the bacterial host cells incorporate the desired non-canonical amino acid in a residue-specific manner. The desired non-canonical amino acid has to be available in excess during expression of the target protein for optimal incorporation. Since the growth medium is supplemented with the desired non-canonical amino acid, this method is termed supplementation incorporation method (SPI).

For the successful production of a completely labeled target protein, additional requirements must be fulfilled. The most crucial is the efficient uptake of the non-canonical amino acid from the growth medium. Fortunately, the amino acid transport systems are quite promiscuous and are therefore not able to distinguish between chemically and structurally similar amino acids (14). The incorporation propensity of a non-canonical amino acid is certainly highly influenced by its intracellular accumulation. Recently, it has been reported that active uptake of fluorotryptophans in breast cancer cells leaded to 70-fold higher intracellular concentrations, than extracellular ones (21).

However, once in the cell, non-canonical amino acids may participate in intracellular chemical reactions and transformations and these reactions may change their chemical nature and/or transform them into toxic compounds. Therefore, the most favorable features of the desired non-canonical amino acids are, to be metabolically neutral and chemically stable.

Besides the high intracellular accumulation level of the desired amino acid analog, its efficient charging of the analog onto the tRNA, by its AARS, is equally important. Although the AARS are crucial enzymes in the interpretation of the genetic code and are highly specific in cognate amino acid recognition, they are often not capable to distinguish catalytically between similar substrates. This feature is known as substrate promiscuity and is quite widespread among many enzymes (14). It is the key feature for the expansion of the amino acid repertoire by the SPI method and makes it quite simple and straightforward. In particular, their promiscuity enables enzymes to either catalyze one specific reaction with a number of similar substrates or to catalyze a range of chemical reactions depending on the reaction partner (22). Recently, Szostak and coworkers have compiled a list of over 90 amino acid analogs that are substrates for wild type AARSs in an *in vitro* reaction (23, 24) and demonstrated that about 70% of the codons of the genetic code could be reassigned to non-canonical amino acids. Thus, a large number of non-canonical amino acids can successfully be incorporated into proteins, since they resemble the canonical counterparts in terms of shape, size and chemical properties. On the other hand, the catalytic efficiency of these analogs is usually much lower than that of their natural counterparts. For example, already in 1979 A. Fersht and C. Dingwall published that the K_m value of Nle for Met-RS from *B. stearothermophilus* was 100 times higher than that for the canonical amino acid Met (25).

In a typical SPI experiment, the non-canonical amino acid analog ought to accumulate in high concentration in the cell, compared to the canonical amino acid, which should efficiently be exhausted. Therefore, the natural AARS can catalyze tRNA charging of the analog. Although it is far less efficient than for the canonical counterpart, translation of the target protein with the novel amino acid should not be significantly affected. The pre- and post-transfer proofreading mechanisms of AARS, as well as ribosome proofreading are normally bypassed in a successful SPI-experiment (14). The last crucial step in protein synthesis is the protein folding, which is influenced by different factors: van der Waals interactions, hydrogen bonds,

Introduction

hydrophobic interactions, hydration effects of non-polar groups, and salt bridges (26, 27). Fig. 2 represents the basic prerequisites for noncanonical amino acid incorporation.

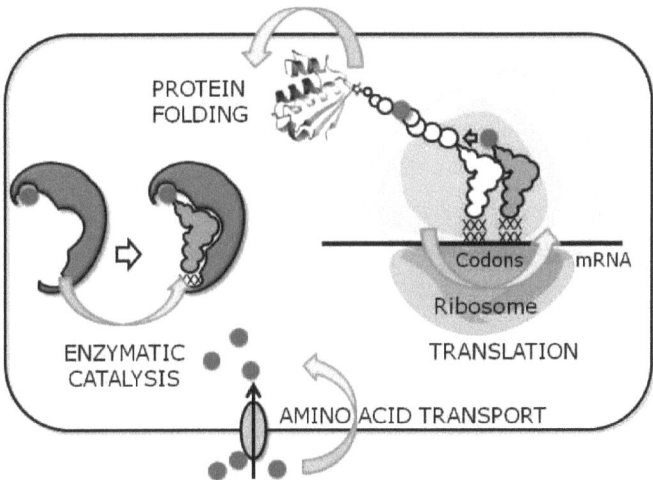

Fig. 2 Basic prerequisites for noncanonical amino acid incorporation. First, the non-canonical amino acid has to be uptaken from the cell. Second, high intracellular concentration of the desired amino acid analog and efficient charging of the analog onto the cognate tRNA is regarded. Third, bypassing of the proofreading mechanisms of AARS and ribosome, and finally proper folding of the target protein. The figure was kindly provided by Dr. B. Wiltschi.

3.5.1 Comparison of different approaches

The basic advantages of the SPI method are first the possibility to direct sense-codon substitution, second the expression and purification of the variant protein at wild type levels and finally the simple and reproducible methodology. A big drawback of this method is that it is not possible to site-specifically exchange one amino acid (all or none exchange manner). Nonetheless, the SPI approach still holds the most promise for large-scale synthesis of proteins with non-canonical amino acids for therapeutic, biomaterial and bioengineering applications.

Introduction

Besides the previously described residue-specific method (SPI method), another approach (site-specific method) was described by Chapeville et al. (28) in 1962. They showed that a misaminoacylateded Ala-tRNA^{Cys}, which was generated by the chemical reduction of Cys-tRNA^{Cys}, is able to incorporate code-specific Ala instead of Cys *in vitro*. This discovery provided a basis for approaches that use chemically or enzymatically aminoacylated tRNAs with changed identity in protein translation. In later methodological developments the focus was on the suppression of termination codons, whose original meaning is suppressed by specific misaminoacylated tRNAs (nonsense suppressor tRNA). In this approach, one of three stop codons is reassigned to any desirable amino acids. In cell-free protein synthesis systems, e.g. amber suppressor tRNAs can be pre-acetylated either chemically, ezymatically or with the help of engineered ribozymes. This allows the incorporation of an unnatural amino acid at the position corresponding to the stop-codon position (29, 30). Though pre-acetylated suppressor tRNAs provide the most general approach for unnatural amino acid incorporation, this approach is limited, due to low product yields that originate from inefficient stop codon suppression and a technically expensive and highly sophisticated experimental setup.

The *in vivo* amber-suppression approach allows the replacement of an amino acid residue by a non-canonical analog at the predefined site in a site-directed manner. Nowadays the common strategy for such site specific insertion of non-canonical amino acids, into recombinant proteins, relies on the read-through of an amber (UAG) stop codon in a mRNA by an amber suppressor tRNA$_{CUA}$ that is acetylated with the desired non-canonical amino acid (15, 31, 32). Like in above described *in vitro* system, the genetic code is expanded *in vivo* by reassigning the amber stop codon to a specific non-canonical amino acid. The key requirement for this approach is that genetically encoded AARS/ tRNA$_{CUA}$ pairs are not interfering with any of the AARS and tRNAs of the endogenous expression system. However, the main bottleneck of this system is the sequence context dependence of the efficiency of the suppression, which further causes relatively low yield of the expressed proteins. In recent years, several other incorporation methods (both *in vitro* and *in vivo*) were developed to incorporate non-canonical amino acids into proteins site-specifically. These are missense suppression (33), frameshift suppression (34, 35), use of

Introduction

rybozymes (36), orthogonal ribosomes (37-39) and even novel DNA base pairs (40-43).

The incorporation of non-canonical amino acids with novel chemical, physical and biological properties into proteins, generates not only a new dimension to protein engineering, but also provides the molecular tools for both, studies of protein folding structure and function, and design of proteins with new or enhanced characteristics. The most commonly used unnatural amino acids are those which are fluorescent or photoactivatable, those that carry heavy atoms or reactive side chains and those that mimic post-translational modifications such as phosphorylation and glycosylation. Expanding the genetic code makes the design and construction of new protein biomolecules possible, which can be used to address challenging problems that are currently not routinely accessible by traditional methods, but find as well a great application field in medicine and technology (15, 44).

3.6 Methionine - a unique amino acid in the genetic code

3.6.1 Properties of methionine

The canonical amino acid methionine (Met) chemically belongs to the sulfur containing amino acids. Biophysically it is classified as nonpolar, and rather modestly hydrophobic (14). However, unlike the side chain of other hydrophobic amino acids, such as Ile, Val or Leu, the Met side chain is rather flexible. This is possibly the reason why Met residues are so often involved in biological processes, where flexibility is required, such as cofactor binding (45). Met is indeed involved in many important processes in the cell. Although sulfur belongs to the same group in the periodic table as oxygen, it is much less electronegative. Thus, this feature affords some of the distinct properties of this canonical amino acid. Moreover, Met plays as well a special role in protein biosynthesis. It is the initiating amino acid in the synthesis of eukaryotic proteins, whereas fMet serves the same function in prokaryotes (see 3.4) (46). Furthermore, it plays an important role in the physiology of the cell. Met functions not only as the methyl donor in biological methylation

Introduction

reaction, but also contributes to the stability of protein structures, with both hydrophobic interaction and hydrogen bonding (46).

Met residues are relatively rare in protein sequences, where they represent just 1.5% of all residues of known protein structures (14). Since Met has a hydrophobic character it is often located in the interior of a protein and just 15% of all Met residues are exposed at the protein surface (14). These surface exposed Mets are very susceptible to oxidation. In some cases, functional changes due to Met oxidation in proteins appear to have pathophysiological significance, as will be discussed later.

Besides Met, there are three other sulfur containing amino acids known in living organisms. These are Cys, homocysteine (Hcy) and taurine (Tau) (Fig. 3). But only Met und Cys are present in protein structures. Cysteine is highly reactive and can form a disulfide bond with another cysteine through the oxidation of the two thiol groups. Although Hcy and Tau are not incorporated into proteins, they play important physiological roles. Hcy is a crucial intermediate in the Met metabolism (see 3.6.3), whereas Tau is known to be present in many animal tissues at higher concentrations than any other amino acid and has been proposed, to act as an antioxidant, an intracellular osmolyte, a membrane stabilizer, and a neurotransmitter (46).

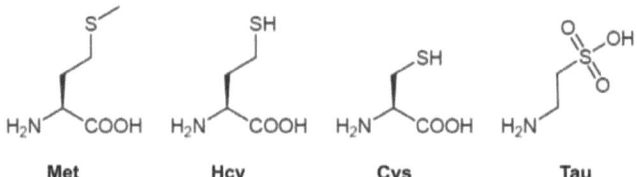

Fig. 3 The four sulfur containing amino acids. Met: Methionine, Hcy: Homocysteine, Cys: Cysteine, Tau: Taurine.

3.6.2 Methionine in the protein synthesis

The first codon translated in all mRNAs is AUG, which codes for Met in eukaryotes and fMet in prokaryotes and eukaryotic organelles. In some cases, the codon GUG (Val) and very rarely the codon UUG (Leu) can be used as well as initiator codons for protein synthesis in bacteria. The initial Met is often cotranslationally removed by a

Introduction

Met aminopeptidase. The tRNA that recognizes the initiation codon differs from the tRNA that carries internal Met residues, although they both recognize the same AUG codon. Presumably, the conformations of these tRNAs are different enough to permit them to be distinguished in the reaction of chain initiation and elongation. However, both tRNAs are aminoacylated by the same MetRS (47). Drabkin and Rajbhandary reported in 1998 that the hydrophobic nature of Met is the key element for being the initiator amino acid. They showed that Val (a highly hydrophobic amino acid) could be used for initiation in eukaryotic cells but that (Glu) glutamine (a polar amino acid) was ineffective (48).

3.6.3 Methionine metabolism

The metabolism of Met is subdivided into *transmethylation*, *remethylation* and *transsulfuration* (46). In the first step of the Met metabolism, Met is activated to S-adenosylmethionine (SAM) by the Met adenosyltransferase (MAT) (Fig. 4/1). SAM was discovered in 1953 by Cantoni (49) as the 'active methionine' and plays a key role in Met metabolism for being the essential biological methyl donor. In the second step, the methyl group of SAM is accepted by an acceptor and S-adenosylhomocysteine (SAH) is formed (Fig. 4/2). In the third step, SAH is hydrolyzed to Hcy and adenosine by the SAH hydrolase (SAHH) (Fig. 4/3). These three steps are summarized as *transmethylation*. Hcy can be methylated back to Met by the Met synthase (MS) and sometimes as well by betaine:homocysteine methyltransferase (BHMT), using 5-methyl-THF and betaine, respectively, as methyl donor. This process is called *remethylation* and is regulated by the need for methyl groups (Fig. 4/4). When the cellular intake of labile methyl groups is high, the need for remethylation is lessened and Hcy is more likely to be catabolized. Catabolism of Met requires the *transsulfuration* process, which involves a two enzyme catalyzed reaction that produces the canonical amino acid Cys. The conversion of Met to Cys is an irreversible process and Cys is further used for glutathione synthesis (Fig. 4/5). An additional important feature of the Met metabolism is its dependence on B vitamin status. In the whole process, four vitamins are involved, three in the *remethylation* pathway and one in *transsulfuration* (46, 50).

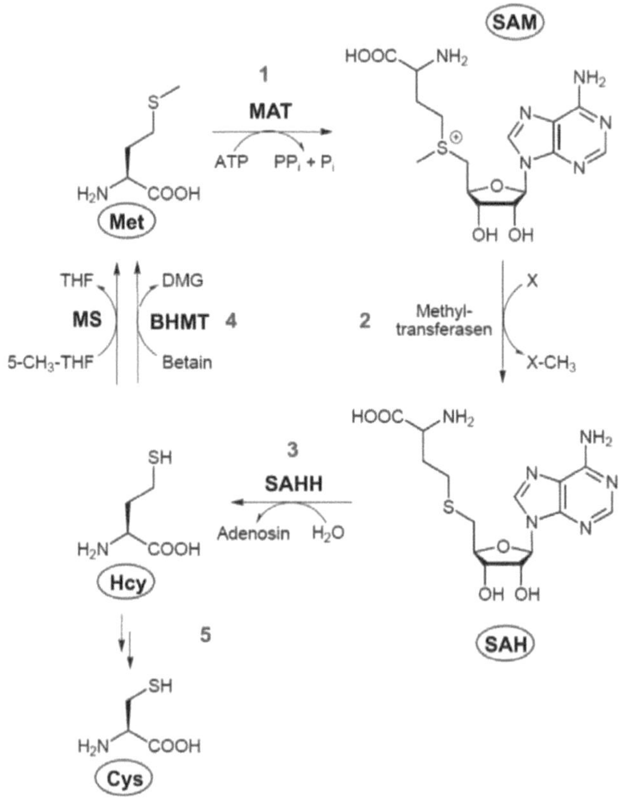

Fig. 4 Met metabolism. MAT: Met adenosyltransferase, SAM: S-adenosylmethionine, SAH: S-adenosylhomocysteine, SAHH: S-adenosylhomocysteine hydrolase, MS: Met synthase, BHMT: betaine:homocysteine methyltransferase, THF: Tetrahydrofolat, DMG: N,N-dimethylglycine

3.6.4 Methionine oxidation

Oxidation of Met is a commonly occurring phenomenon that alters the physicochemical and functional properties of proteins and peptides (51). Met can be readily oxidized to its sulfoxide form, creating a new asymmetric centre and two diastereoisomers, denoted as the S- and the R- form. Most organisms have the potential to express the enzymes, methionine sulfoxide reductase MsrA and MsrB, which can reduce (S)-Met(O) and the (R)-Met(O), respectively. The oxidation degree of Met residues in cells is therefore controlled by the balance between the production of reactive oxygen species (ROS) and the reduction of Met(O) back to Met by the Msr. Further oxidation of Met(O) irreversibly leads to Met sulfone (Met(O_2)), though this occurs to a much lesser extent (Fig. 5). The Met side chain is relatively non polar, compared to many other amino acid side chains and is therefore often found buried within hydrophobic regions of protein structures. Hence, the conversion to its more polar sulfoxide form can have consequential effects on the protein conformation and/or folding. Nonetheless, many important proteins that contain a high number of surface exposed Met residues are known. Naturally, these Met residues are much more readily oxidized than buried ones. Interestingly the oxidation of exposed residues does not appear to have marked effects on the conformation of most proteins. Therefore a significant role of Met in antioxidant defense could be expected, where Met residues act as a ROS scavenger and protect other functionally essential residues from oxidative damage (52). Thus, Met residues might serve as a first defense against ROS damage (53).

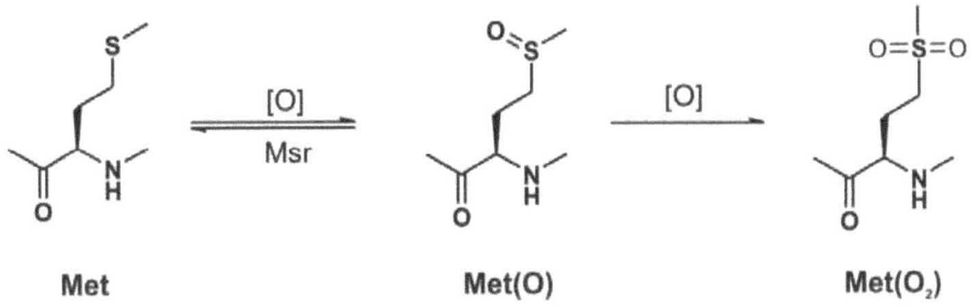

Fig. 5 **Met oxidation to Met(O) and Met(O_2)**. Oxidation of Met to methionine sulfoxide (Met(O)) can be reduced back to Met by methionine sulfoxide reductase (Msr). Further oxidation of Met(O) leads irreversibly to methionine sulfone (Met(O_2)).

When surface exposed Met residues are oxidized by ROS, an oxidation-reduction cycle occurs and Met(O) is reduced by Msr, which itself becomes oxidized. The reduced form is regenerated by thioredoxin, whose oxidized form is then regenerated by thioredoxin reductase using reducing equivalents from NADPH (Fig. 6). This Msr enzymatic system has been directly associated with increased longevity and resistance to oxidative stress in different cell and model organisms (54-56). Furthermore it was shown to participate in the control of intracellular redox homeostasis (54). For example, in transgenic *Drosophila*, overexpression of MsrA

Introduction

extended the mean life span up to 70%, whereas mice without this gene had in average about 40% shorter life spans (57, 58).

Therefore it is not surprising that malfunction of the Msr systems can lead to cellular changes, which result in reduced antioxidant defense, enhanced age-associated diseases involving neurodegeneration and shorter life span (55). Therefore, the functionality of these enzymes is decisive for protein quality and subsequently for life span of every cell and organism.

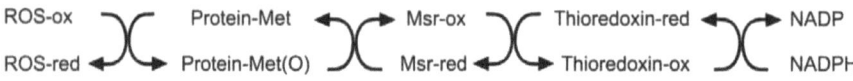

Fig. 6 **Oxidation-reduction cycle.** The enzymatic system that catalyzed the reduction of free and bound MetO utilizes for its action methionine sulfoxide reductase (Msr), thioredoxin and NADPH. ROS: reactive oxygen species, NADP: nicotinamide adenine dinucleotide (oxidized), NADPH: nicotinamide adenine dinucleotide (reduced).

3.7 Proteins and their conformational properties

3.7.1 Protein folding

One of the defining characteristics of living systems is the ability of proteins to fold into their biological functional states. This is one of the most fundamental examples of biological self-assembly (59). The folding and association of nascent or refolding polypeptide chains are known to be autonomous processes. The information required for the formation of the native three dimensional structure of a given protein or protein complex is encoded in the amino acid sequence (Anfinsen rule (60)). Proteins fold into their native conformation through an ordered set of pathways rather than by a random exploration of all the possible conformations until the correct one is stumbled on (12). The structure of a protein is the result of an optimum partitioning of the nonpolar and polar parts of the polypeptide chain between regions of high and low dielectric constant in the solvent environment. While *in vitro* proteins can fold without the presence of accessory proteins, *in vivo* cells contain accessory proteins (molecular chaperones, folding catalysts) or other elaborate strategies, like degradation mechanisms and quality control, to allow proper folding of the polypeptide chains (59). Even after the completion of the initial folding process, proteins are in a dynamic equilibrium between their folded states and a series of more or less unfolded forms (local minima). Thus, a protein can always escape from its correctly folded state into its subsequently misfolded state. In addition when the mechanisms, which are designed to detect and neutralize the effects of such behavior fail, the consequences can be highly disruptive and fatal (61).

3.7.2 Protein misfolding and disease

A broad range of human diseases arise from the failure of a specific protein to adopt or remain in its native functional conformation. Protein misfolding can further lead to aggregation and has been associated with several genetic and environmental factors. The overall mechanism is most likely the destabilization of the native protein conformation and favoring of misfolding and aggregation. Environmental factors that catalyze protein misfolding are metal ions, pathological chaperon proteins, pH or oxidative stress and macromolecular increase in the concentration of the misfolded protein. Many of these alterations are associated with neurodegenerative diseases an aging (62, 63).

The hallmark event in protein misfolded diseases (PMDs) is the change in the secondary and tertiary structure of the native protein that is possibly caused by a primary structural change (e.g. Met oxidation). Therefore, the conformational change may promote the disease either by the gain of a toxic activity or by the loss of the biological function. Additionally, in most of the PMDs, the misfolded protein is rich in β-sheet conformation, which is believed to be the generic state of proteins (64). This means that for many proteins the favorable state is therefore the aggregated one. If this is true, at an infinite time all proteins will be transformed to an aggregated state (65). Thus, it is reasonable to assume that minor stochastically formed modifications of the polypeptide chain induce local α → β structural transition. In this context, Met residues could represent the key structural switches because of their high sensitivity toward oxidation that converts the moderately hydrophobic thioether side chain into the hydrophilic sulfoxide form. Indeed, this redox-controlled Met → Met(O) reaction can induce an α-helix → β-sheet conformational switch in model peptides (66).

Introduction

Although Met is a rare amino acid in proteins (67) there are many important proteins whose activity is altered by Met-oxidation such as 1-antitrypsin calmodulin, fibronectin, cytochrome C, apolipoprotein, chymotrypsin, hemoglobin etc. (52).

An increasing number of human diseases, including Alzheimer's diseases, Parkinson's diseases, Huntington's diseases, spongiform encephalopathies (prion diseases) and late-onset diabetes, are known to be directly associated with deposition of aggregates in tissue (59). Diseases of this type are becoming increasingly prevalent, since the human population generally gets older due to new agricultural, dietary and medical practices. Therefore, most of the PMDs do not result from genetic mutations but develop rather sporadically or are associated with old age. One of the most characteristic features of these diseases is that they give rise to the deposition of proteins in the form of amyloid fibrils and plaques. Several studies shown that amyloid-like fibrils are composed of several protofilaments which consist of hydrogen bonding of β-sheet structures (cross-β conformation) (62, 64). Taken together, it is likely that slight conformational changes result in the formation of a misfolded protein intermediate that becomes stabilized by intermolecular interaction with other molecules and further forms small β-sheet oligomers. That finally leads to amyloid-like fibril formation.

Although all above mentioned neurodegenerative diseases are associated with abnormalities in the folding of different proteins, the molecular pathway, which leads to misfolding and aggregation, as well as the mechanism by which this process might lead to neuronal death, seems to be similar. Thus, these findings provide hope that a common therapeutic strategy can be developed to prevent or treat these defects and diseases.

3.7.3 Oxidative stress and protein damage

Proteins are the major target for oxidants not only because of their abundance in biological systems, but also due to their high rate constants for reaction (53). Depending on the oxidizing species, protein oxidation can lead to either reversible or irreversible protein damage. Radicals and non-radical oxidants can be generated by a wide variety of different processes in biological systems. These range from the deliberate and highly controlled generation of radicals with the active site enzymes to

Introduction

the point of unintended formation of oxidants in cells and tissue. The production of ROS and reactive nitrogen species (RNS) can as well be promoted by different exogeneous or endogenous oxidative stress factors (54). In proteins, almost all amino acids are susceptible to oxidation, but sulfur containing and aromatic amino acids are the most sensitive to oxidative modification. However, specific enzymatic systems have been identified to reduce certain oxidation products, especially for the sulfur containing amino acids, Cys and Met. Furthermore, it is believed that the reversion of Met oxidation may be involved in protein function regulation. Nevertheless, it is well established that oxidation of proteins can have a wide range of downstream consequences. The side chain oxidation of proteins can lead to both, unfolding and conformational changes that can further have consequential effects on their biological function.

Several studies suggested that the oxidation of surface exposed residues have less influence on protein conformational changes than the oxidation of buried ones. Moreover, buried residues are much less rapidly oxidized and additionally require harsher oxidation conditions for being oxidized. In most cases, protein damage is irreversible and the normal fate of these oxidized proteins is their elimination through proteolysis by the intracellular protein degradation pathways. However, heavily oxidized proteins may resist proteolytic attack and form aggregates (68). Additionally, with increasing age the activity of the cellular protein degradation and repair systems decline, which further favors the accumulation of oxidized proteins. Since the steady-state level of oxidized proteins is dependent on the balance between the rate of protein oxidative damage and oxidized protein elimination, it is believed that the age and/or disease related accumulation of oxidized proteins is due to increased protein damage and decreased oxidized protein removal and repair (54, 68).

3.8 Prion disease

Prion protein (PrP) diseases belong to the transmissible spongiform encephalopathies (TSE), which are a group of rapidly progressive, fatal neurodegenerative diseases that affecting both humans and animals. Most TSEs are characterized by long incubation periods and a neuropathological feature of

Introduction

multifocal spongiform changes, astrogliosis, neuronal loss and absence of inflammatory reaction. TSEs in humans include Creuzfeldt-Jakob disease (CJD), Kuru, Gerstmann-Sträussler-Scheinker syndrome (GSS), Fatal Familia Insomnia (FFI) and new variant CJD (nvCJD). In animals it includes scrapie in sheep and goats, transmissible mink encephalopathy in mink, chronic wasting disease in deer and elk, bovine spongiform encephalopathy, exotic ungulate spongiform encephalopathy and feline spongiform encephalopathy in cats, albino tigers and cheetahs (69). Initially, the agent of PrP disease was thought to be a slow virus, but further research indicated that this agent differs significantly from viruses. The *'protein-only'* hypothesis was first enunciated by J. S. Griffith in 1967, where he proposed that the material responsible for disease transmission, was uniquely a protein that has the ability to replicate itself in the body (64). In 1982 Prusiner introduced the term *'prion'* to described the proteinaceous infectious particle (70). Nowadays the best current working definition of a prion is a proteinacious infectious particle that lacks nucleic acid (71).

The cellular prion protein (PrP^C) is encoded by the *PRNP* gene, which directs the synthesis of a 253 residue protein. The first 22 amino acids encode a secretion signal peptide that is cleaved off during its transit through the endoplasmic reticulum (ER) and the Golgi apparatus. It further contains five octapeptide repeats near the amino terminus, two glycosylation sites, one disulfide bridge and additionally a glycosylphosphatidylinositol (GPI) anchor that attaches the protein to the outer surface of the cell membrane. PrP^C is expressed in three different glycosylation forms (mono-, di- and unglycosylated) and is found in the brain and other tissues of healthy individuals (64). Extensive structural studies (72) clearly revealed that in PrP^C the C-terminal region (125-231) adopts an α-helical globular fold with a small two-stranded β-sheet (Fig. 7 A) (73), while the N-terminus (23-124) is mainly unstructured (74). In the human PrP^C 9 Met residues are distributed throughout the structure. Two of them (Met205/206) are buried inside of the hydrophobic core and the others are partly or fully exposed to the outer surface of the protein (Fig. 7). In spite of high sequence conservation in all mammalian species, the specific function of the PrP^C in healthy tissues still remains elusive (75).

Introduction

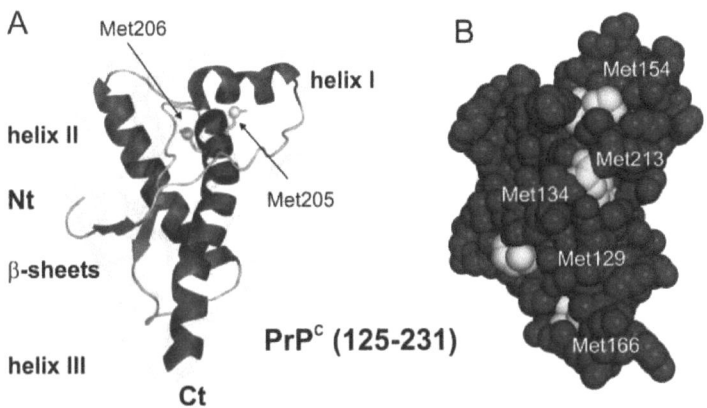

Fig. 7 Three-dimensional structure of human PrP^C(125-231) (PDB accession No. 1QM0) **(A)** Soluble domain of the PrP^C with secondary structure elements and buried Met residues. **(B)** Surface representation with marked Met residues (yellow). Note that except for Met205/206, which are buried inside the PrP^C structure, all other Met residues (including Met109 and Met112, which are part of the unstructured N-terminal domain) are solvent-exposed and therefore in principle susceptible to oxidation.

The key event in PrP diseases is that the normal PrP^C is converted into the scrapie prion protein (PrP^{Sc}), the infectious form of the protein. This procedure is believed to be a posttranslational process, whereby a portion of its α-helical and coil structure is refolded into β-sheet (76) (Fig. 8). This structural transition is accompanied by profound changes in the physicochemical properties of PrP^C. While PrP^C is soluble in nondenaturating detergents, readily digested by proteinase K (PK) and highly α-helical, PrP^{Sc} is insoluble in nondenaturating detergents, partially resistant to PK digestion and has an increased β-sheet content (71). PrP diseases can be of sporadic and inherited and infectious origin (77). Whereas inherited diseases typically arise from mutations in the C-terminal domain, sporadic ones are the most common in humans (85%) (78) and can be assigned to the class of diseases arising from protein conformational disorders such as Alzheimer's and Parkinson's disease (79-81). The infectious process in PrP diseases is believed to be a self-propagating, autocatalytic conformational rearrangement, where recruited and misfolded PrP^C catalyzes the conversion of other PrP^C molecules, which leads to the formation of β-sheet enriched fibrils (82). However, little is known about the initial event in this autocatalytic misfolding cascade. Besides genetic causes (mutations) various environmental factors such as molecular crowding, metal ions, chaperone proteins, membrane lipid composition, pH, and/or oxidative stress have been claimed as responsible (83, 84).

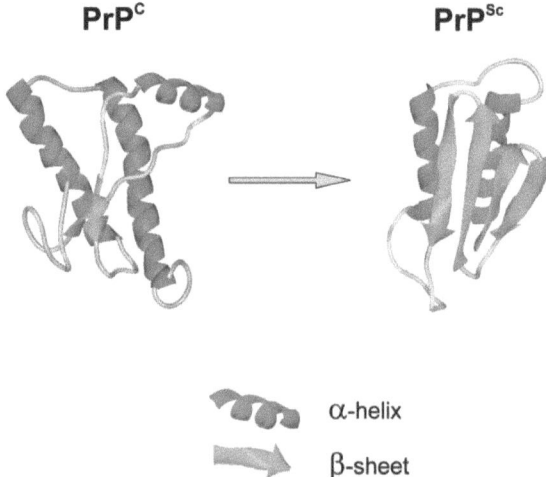

Fig. 8 Structure of PrP^C and speculative structure of PrP^{Sc}. PrP^C shows a high α-helical content, where PrP^{Sc} shows an increase in β-sheet. The picture was adapted from www.stanford.edu/group/virus/prion/normal_rogue.gif.

4 The Goal: Chemical model for prion protein conversion

The central research goal of our group is to generate proteins with properties beyond those of existing ones in the natural biological realm. In this work we aimed at using our synthetic amino acid incorporation methodology to examine the chemical event behind the α → β structural conversion in the recombinant full length human prion protein, since nothing is known about a possible chemical mechanism (85).

Oxidative stress leading to Met oxidation is proposed to be a process capable to promote neurodegenerative disorders, as well as various aging processes by mediating protein misfolding and aggregation. In this context, we reason that a chemical modification of Met residues, such as oxidation, might trigger the structural α → β transition in rhPrP^C, which further initiates the misfolding cascade. To check this hypothesis, we use periodate induced Met oxidation. Thereby, we are not only able to determine the extent of oxidation, but we can also identify crucial Met residues that might be responsible for prion protein transition. In second instance, we generate a suitable model for the structural transition. For an adequate replacement of the amino acid Met, it is necessary that the polarity of the surrogate side chain can be varied in a controlled manner. Therefore, we expand the genetic code for the AUG triplet, by norleucine (Nle) and methoxinine (Mox). These Met analogs are chemically stable and non-invasive and due to their structural similarity to Met, they introduce only the least possible structural perturbation. Importantly, they have opposite polarities. In comparison with Met, Nle is more hydrophobic, whereas Mox (like Met(O)) is a highly hydrophilic (polar) amino acid. It is reasonable to expect that these dramatic differences in physico-chemical properties between Met, Nle and Mox will be fully reflected in the related protein variants.

With this novel tool, we expect to stabilize rhPrP^C, as well as to alter the global folding pattern of rhPrP^C, from a predominantly α-helical structure (like PrP^C) to a structure with predominantly β-sheet content (like it infectious form PrP^{Sc}). In this way, it should be possible to provide a chemical model, which can mimic the α → β structural conversion, not only in prion protein but also in proteins generally important in various neurodegenerative diseases.

5 Material

5.1 Equipment

- Autoclave (Varioklav Dampfsterilisator Typ 500 E; H+P Labortechnik GmbH, Oberschleißheim, Germany)

- Balances (TE1502S; BP211D; Sartorius, Göttingen, Germany; GB2002; PC4400 Delta Range; Mettler-Toledo GmbH, Giessen, Germany)

- CD spectropolarimeter (Jasco J-715 Spectro polarimeter; Temperature control by Peltier FDCD attachment PFD-350S/350L; JASCO International Co., Ltd., Tokyo, Japan)

- Centrifuges (Avanti J-25 Centrifuge; Avanti J-20 XP Centrifuge; Beckmann, Munich, Germany; Centrifuge 5415 C/D; Zentrifuge 3200; Eppendorf, Hamburg, Germany; Universal 32R; Hettich Zentrifugen, Tuttlingen, Germany)

- Centrifuge rotors (JA 25.50; JLA 8.1000; JLA 10.500; Beckmann, Munich, Germany)

- Cuvettes (Hellma 104.002-QS; Hellma 104.002F-QS; Hellma 110-QS; Hellma, Müllheim, Germany)

- Electroporator (Electroporator 1000; Stratagene, La Jolla, CA, USA)

- FPLC instruments (Äktaexplorer; Äktabasic (GE Healthcare, Munich, Germany; columns: Ni-NTA column, GE Healthcare, Munich, Germany; Fractogel EMD-DEAE and Fractogel EMD-SO3 ion exchange columns, Merck KGaA, Darmstadt, Germany))

- French pressure cell (SLM-Aminco)

- HPLC system (C18-RP-HPLC, Waters Alliance 2695 with photodiode array detector 996 and fluorescence detector 2475; column: Waters Xterra RP C18 3.5 µm [2.1x100 mm]; Waters, Eschborn, Germany)

Material

- Incubator (Thermomixer comfort; Thermomixer compact; Eppendorf, Hamburg, Germany; Incubator 3033; GFL, Burgwedel, Germany; Infors-HT AG, Bottmingen, Switzerland)
- Insight Reader (Evotec-Technologies, Germany)
- Magnetic stirrer (MR 3001; Heidolph, Kehlheim, Germany; Ikamag REO; Ika-Combimag REO; IKA® Werke GmbH & Co. KG, Staufen, Germany)
- Mass spectrometer (Pe SCIEX API 165; PerkinElmer Life Sciences, Boston, MA, USA; MicroTOF LC; Bruker Daltonik GmbH, Bremen, Germany; Q-TOF Ultima; Waters, Milford, MA, USA)
- PH meter (MP 220; Mettler-Toledo GmbH, Giessen, Germany)
- Sonifier (Sonifier 450 Macrotip; Branson, St. Louis, MO, USA)
- Sterile bench (Lamin Air HA244GS; Heraeus, Hanau, Germany)
- UV/Vis spectrophotometer (Ultrospec 6300 pro; Amersham Biosciences, Buckinghamshire, UK; UV/VIS spectrometer lambda 19; PerkinElmer Life Sciences, Boston, MA, USA)
- UV lamp (Olympus MT 20 monochromator; Ina-shi, Nagano, Japan)
- Vortex (Reax 1R; Heidolph, Kehlheim, Germany; Vortex Genie 2; Bender & Hobein AG, Zurich, Switzerland)

5.2 Chemicals

Norleucine (Nle) was purchased from Sigma-Aldrich (Taufkirchen, Germany) and methoxinine (Mox) from CBL Patras (Patras, Greece). All other chemicals were from Sigma-Aldrich (Taufkirchen, Germany), Fluka (Buchs, Swiss), Biomol (Hamburg, Germany) and Merk (Darmstadt, Germany) unless stated otherwise. All aqueous buffers and solutions were prepared using bidestilled H_2O (Millipore, Billerica, MA, USA) and autoclaved or sterile filtered if required.

5.3 Buffers and solutions

Resuspension buffer 1	50 mM Tris HCl, pH 8.0; 1 mM $MgCl_2$
Wash buffer	50 mM Tris HCl, pH 8.0; 23% (w/v) sucrose; 0.5 (v/v) Triton X-100; 1 mM EDTA; 1 mM benzamindine
Resuspension buffer 2	8 M urea; 10 mM MOPS, pH 7.0; 50 mM DTT; 1 mM EDTA
Ion exchange buffer 1 (equilibration and wash)	8 M urea; 50 mM Tris HCl, pH 8.0;
Ion exchange buffer 2 (elution)	8 M urea; 50 mM Tris HCl, pH 8.0; 500 mM NaCl
Dilution buffer	8 M urea; 100 mM Tris HCl, pH 8.0; 500 mM NaCl
Ni-NTA buffer 1 (equilibration and wash)	8 M urea; 10 mM MOPS, pH 7.0; 500 mM NaCl
Ni-NTA buffer 2 (elution)	8 M urea; 10 mM MOPS, pH 7.0; 500 mM NaCl; 250 mM imidazole
Refolding buffer	10 mM MES pH6.0
Labeling buffer	20 mM sodium phosphate buffer, pH 7.2; 0.2% sodium dodecylsulfate (SDS, Sigma); 0.1 M $NaHCO_3$
FCS buffer 1	20 mM sodium phosphate buffer, pH 7.2; 0.2% sodium dodecylsulfate (SDS, Sigma)
FCS buffer 2	20 mM sodium phosphate buffer, pH 7.2

Material

Other buffers	
6x SDS-PAGE sample buffer	450 mM Tris·HCl, pH 6.8; 3.6% SDS, 0.2% bromophenol blue; 30% glycerol; 45% β-mercaptoethanol
PAGE Running buffer	0.19 M glycine; 25 mM Tris; 3.5 mM SDS
Coomassie staining solution	0.1% Coomassie; 25% ethanol; 8% acetic acid
Coomassie destaining solution	25% ethanol; 8% acetic acid
Transfer buffer	25 mM Tris pH 8.0; 192 mM glycine
Phosphate buffered saline (PBS)	15 mM Na_2HPO_4; $2H_2O$; 1.8 mM KH_2PO_4; 140 mM NaCl; 2.7 mM KCl; pH 7.4
TPBS	PBS; 0.1% Tween 20

5.4 Polyacrylamide gel electrophoresis (PAGE)

For separation of proteins according to their size first, an appropriate amount of 6x sodium dodecyl sulfate (SDS)-PAGE sample buffer was added to the protein solution and the mix was heated to 95 °C for 5 min.

Second, 17% polyacrylamide gels with 5% stacking gels were prepared and the samples separated by their molecular weight at 120 – 200 V (PAGE Running buffer)

Composition of resolving gel (50 mL)	17% (7.98 mL H_2O; 28.5 mL 30% acryl-bisacrylamide mix; 12.5 mL 1.5 M Tris·HCl, pH 8.8; 0.5 mL 10% SDS; 0.5 mL 10% APS; 0.02 mL TEMED)

Composition of stacking gel (50 mL):	34 mL H$_2$O; 8.5 mL 30% acryl-bisacrylamide mix; 6.25 mL 1.5 M Tris·HCl, pH 6.8; 0.5 mL 10% SDS; 0.5 mL 10% APS; 0.05 mL TEMED

Third, the gels were stained by Coomassie staining solution, and finally destained by Coomassie destaining solution.

5.5 Media

For bacterial growth, fermentation, and protein expression two different media were used: LB medium and New Minimal Medium (NMM) (86).

The components of LB medium, Bacto^{TM} Tryptone and Bacto^{TM} Yeast Extract were purchased from BD Biosciences (San José, CA, USA). 10 g Bacto^{TM} Tryptone, 5 g Bacto^{TM} Yeast Extract, and 10 g NaCl were dissolved in 1 L H$_2$O and autoclaved before usage.

For the incorporation of non-canonical amino acids into protein via SPI method, NMM was used as fermentation medium. For its preparation, first an amino acid mix at pH 7.0 was prepared which contained all amino acids except Met. For this purpose, 0.5 g of each amino acid (as lyophilized powder) were dissolved in 22 mL 1 M KH$_2$PO$_4$, 48 mL 1 M K$_2$HPO$_4$ and H$_2$O *add* 1 L. The solution was sterile filtered before usage. Sterile NMM components were mixed as follows: 7.5 mM (NH$_4$)$_2$SO$_4$; 8.5 mM NaCl; 22.5 mM KH$_2$PO$_4$; 50 mM K$_2$HPO$_4$; 1 mM MgSO$_4$; 20 mM glucose; 100 mL/(L NMM) amino acid mix; 1 µg/(mL NMM) CaCl$_2$; 1 µg/(mL NMM) FeSO$_4$; trace elements 0,01 ng each/(mL NMM) (86); 10 µg/(mL NMM) thiamine; 10 µg/mL NMM biotin and the appropriate antibiotics (100 µg/ml ampicillin and 34 µg/ml chloramphenicol). The stock solutions of Met, Nle and Mox (100 mg/mL) were freshly made and sterile filtered before usage and routinely used for expression and incorporation at a concentration of 5.0 mM without special precautions.

Material

5.6 Enzymes

Lysozyme (application: protein purification)

DNAse I (application: protein purification)

RNAse A (application: protein purification)

Trypsin (application: mass spectrometry; Promega, Madison, WI, USA)

Proteinase K (application: FCS quality control; Merck (Darmstadt, Deutschland)

5.7 Protein molecular weight marker

PageRulerTM Prestained Protein Ladder; Fermentas, St. Leon-Rot, Germany

5.8 Plasmids

5.8.1 pET17b-hPrP(23-231)WT81

The pET17b-hPrP(23-231)WT81 encodes for human PrP23-231 M129 (87), which is under the control of the T7 promoter and provides an ampicillin resistant. The coding sequence of the recombinant full-length human prion protein rhPrP^C(23-231) contains beside the codon for the initial methionine (Met(1)) also serine (Ser(2)) as second amino acid (88). This amino acid composition should facilitate N-terminal Met-excision (89) and subsequently increase protein stability against degradation, according to the N-End Rules (90). Therefore, the gene rhPrP^C(23-231) encodes for 211 amino acids, whereas in the purified proteins N-terminal Met residue is expected to be excised.

5.8.2 pRIL

The chloramphenicol resistant rare codon plasmid, pRIL (Stratagene, La Jolla, CA, USA) is a derivative of pACYC184 plasmid and contains tRNA-genes for the rare Arg, Ile, and Leu codons (AGG, AGA, AUA, and CUA, respectively).

5.9 Antibiotics

Ampicillin (working concentration 100 µg/mL)

Chloramphenicol (working concentration 34 µg/mL)

5.10 Bacterial strains

Met-auxotrophic *Escherichia coli* strain B834(DE3) (F-ompT hsdSD(rD- mD-) gal dcm metE λ(DE3)) (Novagen Merck Chemicals Ltd., Nottingham, UK).

5.11 Fluorescent dyes

Alexa488 (Alexa Fluor-488-O-Succinimidylester) Molecular Probes (Eugene, USA); MW = 643.26 Da

Alexa647 (Alexa Fluor-647-O-Succinimidylester) Molecular Probes (Eugene, USA); MW = 1300 Da

5.12 Antibodies

Primary antibody: mAb 385 (monoclonal mouse antibody, anti-hPrP octarepeats), (ZNP, LMU, Martinsried, Germany)

Secondary antibody: goat anti-mouse IgG coupled to alkaline phosphatase (Dianova, Hamburg, Germany)

5.13 Software

- Except standard software, Origin 6.1G (OriginLab Corporation, Northampton, MA, USA) was applied for data analysis.
- The calculated composition of the different secondary structures of the two peptide analogues, Nle and Mox, of the Dado-Gellmann model peptide was computed with the CONTIN software (Reference SMP56)/CDPro Analysis.

Material

- Orbitrap ESI-MS/MS data analysis was performed with the 1.0.9.19 MaxQuant software (Matrix Science, London, UK), supported by Mascot as the database search engine for peptide identifications.

- The fluorescence data were analyzed by correlation analysis using the FCSPPEvaluation software version 2.0 (Evotec OAI, Germany) as described (91). Two-colour cross-correlation amplitudes G(0) were determined using the same software.

6 Methods

6.1 Microbiological methods

6.1.1 Production of electrocompetent cells

First cells were incubated overnight at 37 °C in 5 mL LB medium without antibiotics. On the following day, 1 L LB medium without antibiotics was inoculated using the 5 mL overnight culture and incubated at 37 °C and 220 rpm until the OD_{600} of the cell culture was between 0.6 and 0.8. The cells were harvested and washed twice with ice-cold 10% glycerol, subsequently. Finally the cell pellet was resuspended in 5 mL 10% glycerol, aliquots à 100 µL were frozen in liquid nitrogen, and kept at -80 °C until further use.

6.1.2 Transformation

Bacterial cells were transformed by electroporation. For this purpose, a 100 µL competent cell aliquot of the appropriate cell strain was mixed with ~1 µg of the required plasmid in an electroporation cuvette. The cell suspension was shocked by 1650 V, mixed with 1 mL chilled LB medium, and transferred to a sterile eppendorf cup, subsequently. In the following, the cells were incubated at 37 °C and 800 rpm for 1 h. Finally, the cells were distributed on agarose gel plates with the appropriate antibiotics.

On the next day, single colonies were transferred to 5 mL of LB medium with the appropriate antibiotics and incubated overnight at 37 °C and 220 rpm.

6.1.3 Limitation test

For the incorporation of non-canonical amino acids into proteins during protein expression it is absolutely necessary that the applied cell strain is auxotrophic for the respective canonical amino acid. Therefore, prior to the incorporation experiment, it is essential to prove the cell´s auxotrophy and determine the optimal amino acid concentration for biomass production by a limitation test. For this purpose, 5 mL NMM with appropriate antibiotics and different concentrations of L-Met were inoculated with 50 µL of pre-culture. These suspensions were incubated overnight at

Methods

37 °C and 220 rpm and on the next day, cell growth was checked by measurement of OD_{600}. The amino acid concentration which permitted cell growth up to an OD_{600} between 0.6 and 0.8 was utilized as limiting concentration for biomass production for incorporation experiments.

6.1.4 Expression test

To select the best expression clone, 1 mL of each overnight cell culture was transferred to a sterile eppendorf cup and the protein expression was induced by adding 1 mM IPTG. The cell cultures were incubated at 800 rpm and 30 °C. In parallel, a non-induced sample (without addition of IPTG) was prepared.

After incubation, the cells were first harvested, second resuspended in 50 µL H_2O, and finally heated to 95 °C for 5 min after addition of 10 µL 6x SDS sample buffer. Afterwards, protein expression was checked by SDS-PAGE and the best expressing clone was selected for subsequent protein expression.

6.1.5 Protein fermentation and expression of full length $rhPrP^C$

All fermentation and expression experiments were performed in the above described synthetic medium (NMM). Transformed cells were grown in 5 mL LB-medium overnight with 100 µg/mL ampicillin and 34 µg/mL chloamphenicol. 100 µL of the overnight culture were used for the inoculation of 1 L NMM. Cells were grown to mid-log phase (OD_{600} 0.6-0.8) in NMM containing limiting amounts (0.04 mM) of L-Met. Protein expression in Met starved cells was induced with 1 mM isopropyl-β-D-1-thiogalactopyranoside (IPTG; Applichem, Darmstadt, Germany) and 5.0 mM L-Met, D, L-Nle or L-Mox was concomitantly added. Proteins expression was performed for 4 hours to overnight at 30 °C with aeration achieved by vigorous shaking (220 rpm).

After cell harvest, cell pellets were resuspended in resuspension buffer 1 and frozen at -20 °C until further purification.

6.1.6 Protein purification of full length $rhPrP^C$

Proteins were expressed in inclusion bodies. The isolation and refolding protocol was identical for all variants. Based on the published protocols (92) cells pellets were thaw and disrupted in a French pressure cell (15000 psi; SLM-Aminco). Lysozym,

Methods

DNase and RNase were added to the disrupted cells and were vigorously shaked for 20 min at 30°C. The inclusion bodies were sedimented by centrifugation (25 min, 49000g, 4 °C) and washed twice with wash buffer. The pellet was dissolved in resuspension buffer 2 and was centrifuged (30 min, 130000g, 22°C). The supernatant was applied to linked Fractogel EMD-DEAE and Fractogel EMD-SO3 ion exchange columns (Merck KGaA, Darmstadt, Germany), 25 mL and 15 mL respectively, which were equilibrated in ion exchange buffer 1. Proteins were eluted from the Fractogel EMD-SO$_3$ column with ion exchange buffer 2. All fractions containing solubilized rhPrP^{C}s were pooled and diluted in dilution buffer to a protein concentration of 0.05 mg/mL. Subsequently, the CuSO$_4$ was added to a final concentration of 2 µM, and the solution was stirred for 4 h at room temperature. As a result, the formation of the single disulfide bond takes place. The oxidation reaction was quenched by addition of 1 mM EDTA. Excessive EDTA was bound with the addition of 1 mM NiAc and pH was adjusted to about 6.8 with 1 M HCl. The oxidized PrP^{C}s were applied to 5 mL Chelating Sepharose Fast Flow (GE Healthcare, Germany), pre-charged with NiAc according to the manufacturer's recommendations and pre-equilibrated with Ni-NTA buffer 1. The bound proteins were eluted with Ni-NTA buffer 2. Protein containing fractions were pooled and refolding was done in 50-fold volume in refolding buffer at 4 °C overnight. Refolded proteins were concentrated and stored at 4 °C.

6.1.7 Protein purity and concentration determination

The purity of the recombinant protein samples was demonstrated by SDS-PAGE, HPLC-profile analysis and electrospray mass spectra analysis (ESI-MS). Concentrations of rhPrP^C samples were determined using the UV/VIS spectrometer lambda 19 (PerkinElmer Life Sciences, Boston, MA, USA) assuming an extinction coefficient of 57932.5 M^{-1}cm^{-1} as calculated from the amino acid composition using SwissProtParam software tools (Lambert-Beer-Equation $A = \frac{log \, i_0}{ln \, g^l} = c \cdot d \cdot \varepsilon$,

A = absorbance; ε = molar extinction coefficient; c = concentration; d = path length). Protein concentrations were calculated as described elsewhere (93).

6.2 Spectroscopy and spectrometry

6.2.1 Mass spectrometry (Orbitrap ESI-MS/MS)

6.2.1.1 In-solution Digestion.

Samples were diluted 4 X in a 6 M urea/2 M thiourea (in 10 mM Hepes, pH 8.0) to reach denaturing conditions. To reduce disulfide bonds 100 mM DTT was added to a final concentration of 10 mM in the protein solutions and incubated for 45 min at RT. The resulting free thiol (-SH) groups were subsequently alkylated with iodoacetamide (55 mM final concentration) for 30 min at RT in the dark. The solutions were then digested with Lys-C (Wako) (1 µg/50 µg sample protein for 3 h at RT) diluted 4X with 50 mM NH_4HCO_3, and digested with Trypsin (Promega) (1 µg/50 µg sample protein for 16 h at 37 °C) (94). Peptide mixtures were then desalted by using Stop and go extraction (STAGE) tips and the eluted peptides used for mass spectrometric analysis.

6.2.1.2 NanoLC-MS/MS

All digested peptide mixtures were separated by on-line nanoLC and analyzed by electrospray tandem mass spectrometry. The experiments were performed on an Agilent 1200 nanoflow system connected to an LTQ Orbitrap mass spectrometer (Thermo Electron, Bremen, Germany) equipped with a nanoelectrospray ion source (Proxeon Biosystems, Odense, Denmark). Binding and chromatographic separation of the peptides took place in a 15 cm fused silica emitter (75 µm inner diameter from Proxeon Biosystems, Odense, Denmark) in-house packed with reversed-phase ReproSil-Pur C18-AQ 3 µm resin (Dr. Maisch GmbH, Ammerbuch-Entringen, Germany). Peptides were eluted with a 140 min linear gradient of 98% solvent A (0.5% acetic acid (Fluka) in H_2O) to 50% solvent B (80% acetonitrile (Merck) and 0.5% acetic acid in H_2O). The precursor ion spectra were acquired in the Orbitrap analyzer (m/z 300–1800, R = 60,000, and ion accumulation to a target value of 1,000,000), and the five most intense ions were fragmented and recorded in the ion trap. The lock mass option enabled accurate mass measurement in both MS and Orbitrap MS/MS mode as described previously (95). Target ions already selected for MS/MS were dynamically excluded for 90 s.

Methods

6.2.1.3 Peptide Identification via MASCOT Search

The data analysis was performed with the MaxQuant software as described (96) supported by Mascot as the database search engine for peptide identifications. Peaks in MS scans were determined as three-dimensional hills in the mass-retention time plane. MS/MS peak lists were filtered to contain at most six peaks per 100 Da interval and searched by Mascot (Matrix Science) against a forward and reversed version of the IPI Human database. The initial mass tolerance in MS mode was set to 7 ppm. and MS/MS mass tolerance was 0.5 Da. Cysteine carbamidomethylation was searched as a fixed modification, whereas oxidized Met, Tyr, Trp and His were searched as variable modifications.

6.2.2 Electrospray mass spectrometry (ESI-MS)

The extent of replacement of the native Met residues by Nle or Mox was confirmed by high resolution mass analysis using liquid chromatography (LC) coupled with the electrospray ionization-time of flight (ESI-TOF) mass spectrometer MicroTOF-LC from Bruker Daltonics (Bremen, Germany). Samples were separated by Symmetry C4 column (Waters, Eschborn, Germany) with a flow rate of 250 µL/min and 15 min linear gradient from 20% - 80% acetonitrile/0.05% trifluoroacetic acid.

6.2.3 Circular dichroism (CD) spectroscopy and melting curves

The far-UV CD spectra of full-length rhPrP^{C}s were recorded on a dichrograph JASCO J-715 (JASCO International Co., Ltd., Tokio, Japan). All spectra are averages of four scans and are reported as mean residue molar ellipticity ($[\theta]_R$) in degrees x cm^2 x dmol^{-1}. Quartz cells (110-QS Hellma) of 0.1 cm optical path length and protein concentrations of 8.73 µM were used. Ellipticity changes were recorded between 200 nm and 260 nm at 37 °C in 10 mM MES pH 6.0. Melting curves were determined by monitoring the changes in dichroic intensity at 222 nm in function of temperature increase. Thermal denaturation experiments were performed in the range from 4 °C to 95 °C, with a heating rate of 30 °C h^{-1}. Temperature was controlled by a JASCO Peltier type FDCD attachment (model PFD-350S/350L; JASCO International Co., Ltd., Tokyo, Japan). The fractions of unfolded protein were calculated from the corresponding ellipticity data as well as the midpoint of denaturation (melting temperature or T_m value) (92).

6.2.4 Quantification of individual Met residue oxidation in rhPrP^C

The oxidation of Met-rhPrP^C was performed at 0.4 mM protein concentration with 0.5, 5 and 25 equiv. sodium periodate in 10 mM MES, pH 6.0, at 0 °C and overnight, according to well established protocols (97). After tryptic digestion the peptide mixture was analyzed by ESI-MS. Samples were separated by Symmetry Varian Pursuit XRs ultra C18 column (Varian, Germany) with a flow rate of 250 µL/min and 80 min linear gradient from 30% - 90% acetonitrile/0.05% trifluoroacetic acid.

6.3 Peptide synthesis

The two peptides were kindly provided by the Microchemistry Core Facility (MPI of Biochemistry, Mrs. S. Andric). The peptides were synthesized, using the 433A Peptide Synthesizer (Applied Biosystems, USA). After synthesis a test cleavage from the resin was done, which was further analyzed using an analytical RP-HPLC (Waters, Eschborn, Germany), using an EC 125/4 Nucleosil 100-5 C8 colum (Macherey Nagel, Düren, Germany), with a flow rate of 1.5 mL/min. The used gradient was from 100% A (2% H_3PO_4 (95%) and acetonitrile (5%)) to 100% B (2% H_3PO_4 (10%) and acetonitrile (90%)) in 28 min. In addition ESI-MS was performed. Subsequently, remain of the synthesized peptides were cleaved from the resin with 96% triflouracetic acid, 2% triisopropylsilane and 2% water. Afterwards the peptides were precipitated using n-hexane/*tert*-butylmethylether (1:2, v/v). The precipitated Nle-peptide (83 g) was dissolved in 10% acetic acid and 90% water. The Mox-peptide (55 g), which was less soluble, was dissolved in 5% triflouracetic acid, 25% acetonitrile and 70% water. The purification of the peptide solutions was done using a preparative RP-HPLC (Gilson Abimed, Langenfeld, Germany). For that purpose an EC 125/4 Nucleosil 100-5 C18 column (Macherey Nagel, Düren, Germany) with a flow rate of 1.2 mL/min and a gradient of 10% - 90% buffer B (buffer A (0.1% triflouracetic acid in water) and buffer B (0.08% triflouracetic acid in acetonitrile) in 90 min was used. The correct mass of the peptides were confirmed

Methods

using ESI-MS. Fractions containing the peptides with the correct mass were pooled and lyophilized, giving a yield of 10 mg each.

6.4 FCS/SIFT measurements and analysis

The theoretical concept of FCS (fluorescence correlation spectroscopy) and SIFT (scanning for intensely fluorescent targets) are explained in detail in Schwille (91) and Bieschke et al. (98) respectively.

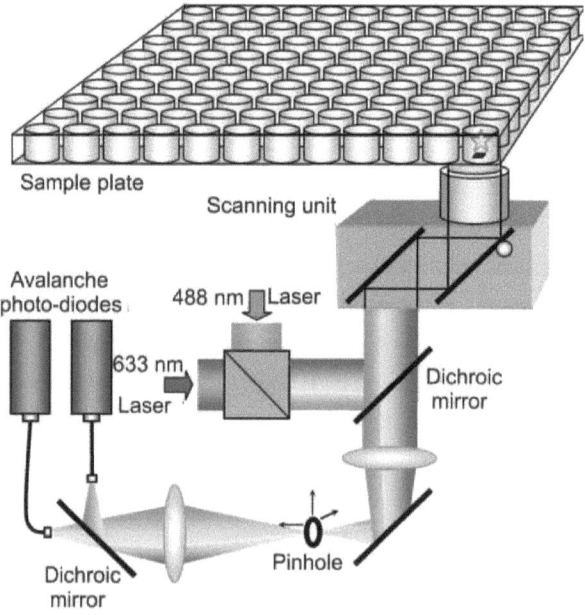

Fig. 9 **Measuring systems of a dual color FCS reader with a scanning unit for SIFT measurements and sample plate.** By using a dichroic mirror (reflect light just of a special wavelength) and a confocal microscope, the laser light is focused into the sample. The fluorescence light passes the dichroic mirror and via a pinhole it finds its way to the photodetectors, the so called avalanche photo-diodes. The picture was kindly provided by Prof. Dr. Armin Giese.

FCS is based on the analysis of fluctuations in fluorescence caused by the diffusion of fluorescently labeled molecules at nanomolar to picomolar concentrations

Methods

through an open detection volume of approximately 1 fL defined by a focused laser beam. When molecules labeled with two different fluorophores form complexes, the amount of complex formation can be easily monitored and quantified by cross-correlation analysis in a dual-color setup (91, 98-101).

FCS - using a stationary focus - and SIFT - using a mobile focus - measurements were carried out (Insight Reader, Evotec-Technologies, Germany) with dual-color excitation at 488 nm and 633 nm, using a 40 x 1.2 NA microscope objective (Olympus, Japan) and a pinhole diameter of 70 µm at FIDA setting (Fig. 9). Excitation power was 200 µW at 488 nm and 300 µW at 633 nm. Measurement time was 5 x 10 sec. For scanned measurements, scanning parameters were set to 100 µm scan path length, 50 Hz beamscanner frequency, and 2000 µm positioning table movement. This is equivalent to approximately 10 mm/s scanning speed. The temperature for all measurements was 20-25 °C. The fluorescence data were analyzed by correlation analysis using the FCSPPEvaluation software version 2.0 (Evotec OAI, Germany) as described (91). Two-color cross-correlation amplitudes G(0) were determined using the same software. Evaluation of SIFT data in two-dimensional intensity distribution histograms was performed as described previously (98).

To quantify aggregate formation, cross-correlation analysis was used, because dual-color cross-correlation analysis allows more sensitive detection and quantification than single-color auto-correlation analysis. Moreover, cross-correlation analysis can be combined with scanning, which further improves sensitivity of aggregate detection (99).

6.4.1 *Fluorescent labeling of rhPrP^C*

Protein labeling was performed with the amino-reactive fluorescent dyes Alexa Fluor-488- and Alexa Fluor-647-O-succinimidylester (Molecular Probes, USA), respectively. The fluorescent dyes (2 mg/mL in DMSO) were added to a solution of 8.73 µM (10 µg/50 µL labeling buffer) rhPrP^C at a molar ratio of 3:1. After incubation for 2 h at room temperature in darkness, unbound fluorophores were separated by two filtration steps in Microspin columns filled with Sephadex G25 (Pharmacia Biotech, Sweden), equilibrated with 20 mM sodium phosphate buffer (pH 7.2)

Methods

containing 0.2% sodium dodecylsulfate (SDS, Sigma) (fraction A) (Fig. 10). Both columns were washed again with 50 μL FCS buffer 1, to collect not eluted rhPrP^C (fraction B) (Fig. 7). Removal of unbound dye molecules, labeling efficiency and pre-aggregation were confirmed by FCS measurements on an Insight Reader (Evotec-Technologies, Germany). Fractions with more than 10% of unbound dye molecules or aggregates that were not soluble in 0.2% SDS were discarded. After quality control the stock solutions were aliquoted and stored at -80 °C.

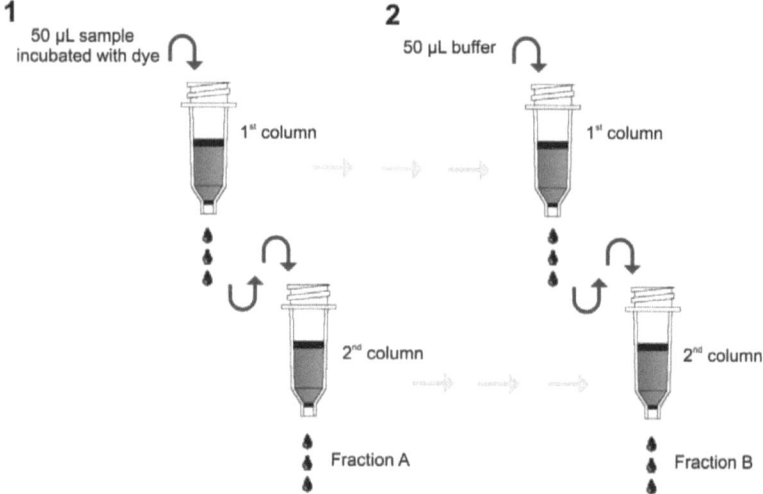

Fig. 10 Flow chart of the labeling procedure.

6.4.2 Quality control of labeled rhPrP^C

The quality control of the rhPrP^C labeling, of fraction A and B, was done with a 1:1000 dilution of the stock solutions, which were measured by FCS and SIFT. By using the two components fit of the FCS measuring, the number of free dye molecules and labeled rhPrP^C, as well as the diffusion time of both was determined. This was done by entering the known diffusion time of Alexa 488 and Alexa 647, respectively. For these fixed values the corresponding percentage out of the total signal could be appraised and therefore the diffusion times of the labeled proteins as well as the number of particles (N) was calculated by the evaluation software FCS+plus_Eval 2.0 (Evotec OAI, Germany). On the basis of the proteins diffusion

Methods

times it was possible to show, if the proteins were degradated while labeling procedure. Samples with incorrect diffusion times were discarded. The theoretical diffusion time (τ) follows the equation below:

$$\tau_{Protein} = \sqrt[3]{\frac{MW\ Protein}{MW\ Alexa}} \times \tau_{Alexa}$$

By using the SIFT measurement, it was possible to look for insoluble aggregates in 0.2% SDS. Samples that contained aggregates or diffusion times which differ from the estimated ones were discharged. An additional quality control was done by checking the amount of dye molecules pro rhPrP^C molecule. For this purpose 0.1 mg/mL proteinase K was added to the different protein samples (1:1000 dilutions) which were further incubated for one hour at 37 °C. After one hour the digest was observed using FCS. In this particular case, two parameters were particularly important, the diffusion time of the digested protein and the N value. Because by dividing the number of particles before digestion and after digestion, you can calculate the number of dye molecules pro rhPrP^C molecule. The optimal labeling ratio ($N_{after\ digest}/N_{before\ digest}$) should be over 1 and below 2 (see Table 1 and Table 2). Although this labeling quality control shows the labeling ratio, it is not possible to detect unlabeled protein. Thus, it is necessary to proof this by western blot.

N	rhPrP^C-Met (1:1000)	rhPrP^C-Nle (1:1000)	rhPrP^C-Mox (1:1000)	Fraction
Alexa 488	1.80	1.41	4.49	A
Alexa 647	5.47	2.59	8.45	A

Table 1   N values before PK digest. Measurements were done with 1:1000 dilutions of all samples.

N	rhPrP^C-Met (1:1000)	rhPrP^C-Nle (1:1000)	rhPrP^C-Mox (1:1000)	Fraction
Alexa 488	3.02	2.43	7.97	A
Alexa 647	8.55	4.52	14.33	A

Table 2 **N values after PK digest.** Measurements were done with 1:1000 dilutions of all samples.

6.4.3 Western blot

A SDS-PAGE was done with equal amounts of each fraction and after electrophoresis, the separated molecules were transferred onto a Whatman Protan nitrocellulose transfer membrane (Dassel, Germany) using a semi-dry blot (MPI, Martinsried, Germany). Before blotting, the membrane and the filter papers were soaked in Transfer buffer. The 'sandwich' construct was blotted with 200 mA for 2 hours. Next, the membrane was blocked in TPBS plus 3% BSA for one hour (RT) or overnight (4 °C), to prevent any nonspecific binding of antibodies to the surface of the membrane. After the incubation with the first antibody (mAb 385; 1:200) for one hour at RT the membrane was washed three times for about 10 min with TPBS buffer. The membrane was then incubated for one hour at RT with the secondary antibody (1:10000), conjugated with alkaline phosphatase for chemiluminescent detection. Before detection, the membrane was again washed three times for 10 minutes at RT to remove any unbound antibody. Enzymatic activity was visualized with the Pierce ECL (Thermo Fisher Scientific, Germany) substrate as described by the manufacturers.

6.4.4 Aggregation assay

For the *in vitro* aggregation assay, the stock solutions, labeled as described above, were diluted in two steps. The stock solutions of rhPrP^C-Alexa 488 were first diluted to N = 20, whereas the solutions of rhPrP^C-Alexa 647 were adjusted to N = 30 (the different N values correspond to a real stoichiometric ratio of 1:1). This dilution step was done, to get a balanced ratio of green and red labeled rhPrP^C. The second dilution step was a 1:16 or 1:10 dilution that was performed in buffer containing 0% or 0.2% SDS, to obtain final SDS concentrations of 0.0125% and 0.2% SDS,

Methods

respectively (Fig. 11). The final rhPrP^C concentration was about 5 nM. Experiments were done in 96-well-plates with a cover slide bottom (Evotec-Technologies, Germany). To reduce evaporation, the 96-well-plates were sealed with adhesive film. Aggregation was monitored for about 12 hours using four or eight parallel and identical samples for each experimental group.

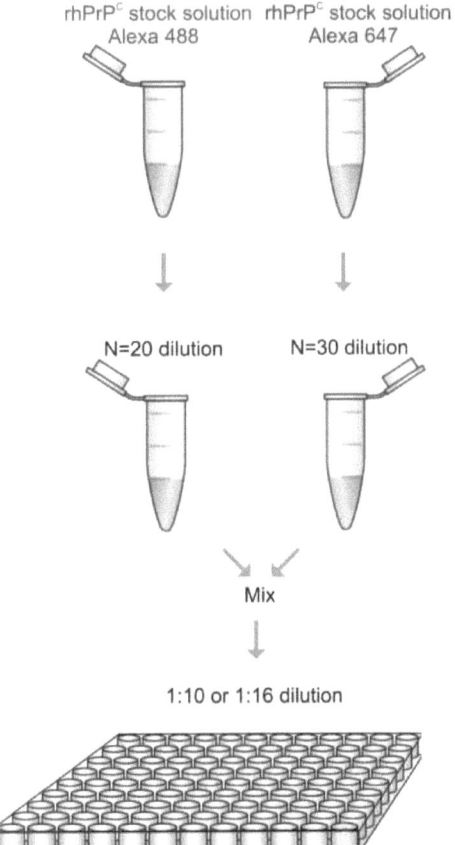

Fig. 11 Flow chart of dilution steps for aggregation measurements.

6.4.5 Aggregation assay in different periodate concentrations

The experimental setup was performed as described before. The only difference was that different volumes of sodium periodate (NaIO$_4$) stock solution (100 µM and

Methods

50 mM) were pipetted in order to give final concentration of 2 µM, 200 µM and 10 mM $NaIO_4$ in the examined samples during the first dilution step. The samples were incubated for about 12 hours and further diluted for measurement as described before.

7 Results

7.1 Expression and purification of the rhPrP^C model protein

For our studies we used the pET17b-hPrP(23-231)WT81 plasmid that encodes for the 211 amino acid long human PrP23-231 M129 (87) (Fig. 12).

MSKKRPKPGG	WNTGGSRYPG	QGSPGGNRYP	PQGGGGWGQP
HGGGWGQPHG	GGWGQPHGGG	WGQPHGGGWG	QGGGTHSQWN
KPSKPKTNMK	HMAGAAAAGA	VVGGLGGYML	GSAMSRPIIH
FGSDYEDRYY	RENMHRYPNQ	VYYRPMDEYS	NQNNFVHDCV
NITIKQHTVT	TTTKGENFTE	TDVKMMERVV	EQMCITQYER
ESQAYYQRGS	S		

Fig. 12 Amino acid sequence of PrP23-231 M129.

To gain better protein expression, due to the codon usage differences, we used an additional plasimd pRIL in co-expression experiment (see 6.1.2). After expression, the protein was deposited as inclusion bodies in the cytoplasm. They were first dissolved in 8 M urea and therefore the whole purification was performed under these conditions. We used for the first purification steps a Fractogel EMD-DEAE in combination with a Fractogel EMD-SO3 ion exchange column (Merck KGaA, Darmstadt, Germany). The fractions which contained the protein were pooled and oxidized. Further purification was done by using a Ni-NTA column (GE Healthcare, Germany). The purified protein was refolded overnight at 4 °C (see 6.1.6). Fig. 13 shows Coomassie stained SDS-PAGE gels from the different purification steps.

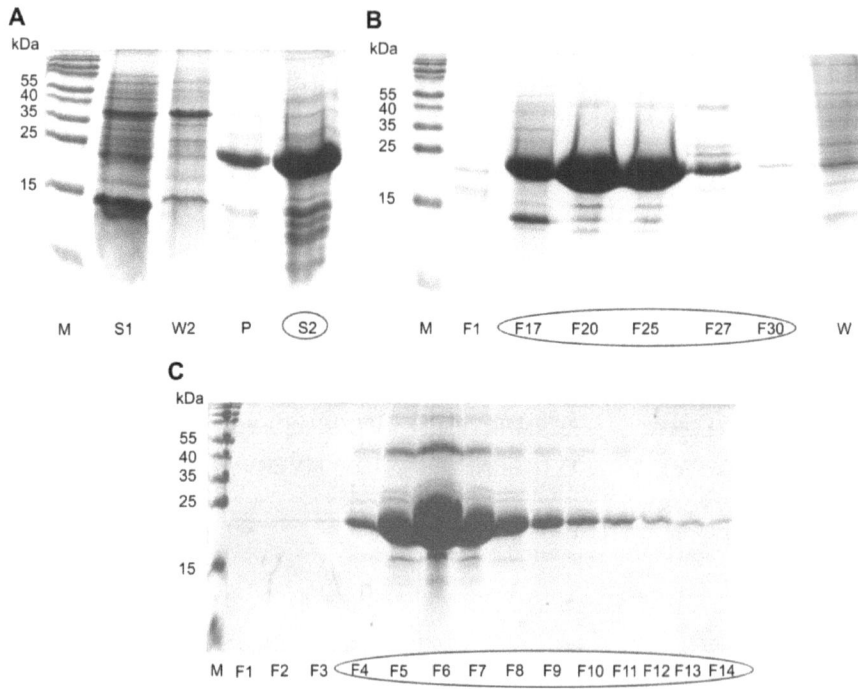

Fig. 13 **Coomassie stained SDS-PAGE gels of different purification steps.** (**A**) M: marker; S1: supernatant 1; W2: wash 2; P: undissolved pellet; S2: supernatant 2 (Met-PrP^C in urea). (**B**) M: marker; F: fraction; W: wash. (**C**) M: marker; F: fraction. The cycles highlight the fractions, which were further used.

The Coomassie stained SDS-PAGE gels (Fig. 13) clearly demonstrate the sufficient protein purity and yield. In addition, the correct mass of the protein sample has been proven by ESI-MS. The theoretical molecular weight of Met-rhPrP^C is 22919.2 Da, which is in perfect agreement to the mass found by mass spectrometry (22919.2 Da) (Fig. 14). The heterogeneity of the protein sample, seen in the mass spectra, is probably due to metal-protein adducts, most probably generated by unspecific binding of Na$^+$ ions from buffer.

Results

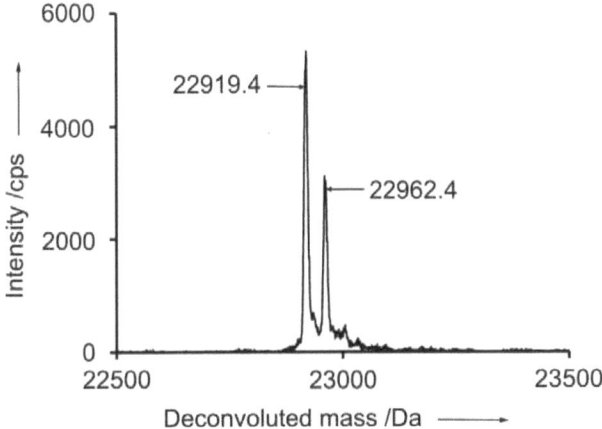

Fig. 14 Deconvoluted mass spectrum of Met-rhPrP^C. The predominant peak (22919.4 Da) corresponds to the mass of Met-rhPrP^C. The accompanying peak most probably originates from unspecifically bound Na$^+$-adducts from buffer.

7.2 Sodium periodate induced aggregation of Met-rhPrP^C

The *in vitro* aggregation tendency of Met-rhPrP^C was monitored using the well established *de novo* SDS-dependent *in vitro* aggregation assay (87, 101), based on confocal single molecule analysis with fluorescence cross-correlation spectroscopy analysis (91, 98). In 0.2% SDS, rhPrP^C exhibits a high α-helical content and does not form aggregates. By diluting the SDS concentration to 0.0125%, a significant rise in cross-correlation amplitude indicates aggregation of rhPrP^C.

To study the effect of Met oxidation on the protein aggregation propensity, comparative aggregation assays were performed under different oxidative conditions. Peroxide (H_2O_2) is routinely used in oxidation studies of many proteins including PrP^{C}s from various organisms (102, 103). However, in our hands sodium periodate (NaIO$_4$) proved to be the best oxidant to generate reliable and reproducible data. Each dilution of the protein was treated with a molar excess of NaIO$_4$ (2 µM, 200 µM and 10 mM). The high excess of oxidant served to overcome the drastically reduced reaction rates at the low protein concentration necessary for FCS/SIFT

Results

measurements. As shown in Fig. 15, with an increase of the oxidant concentration a significantly enhanced aggregation of Met-rhPrP^C can be observed.

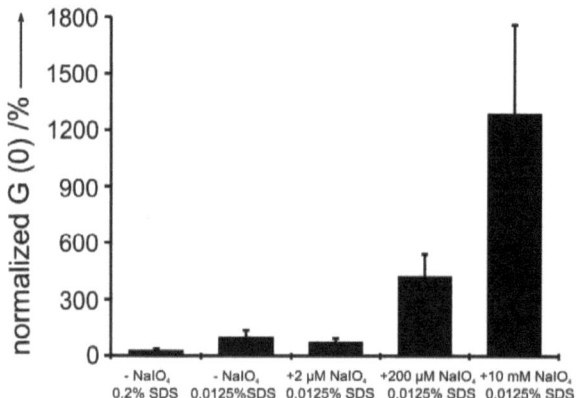

Fig. 15 Periodate dependent aggregation of Met-rhPrP^C determined by cross-correlation amplitude G(0). Each data point represents the mean of four parallel samples. G(0) of Met-rhPrP^C without NaIO$_4$ was set to 100%, to compare the G(0) values of Met-rhPrP^C in different NaIO$_4$ concentrations.

7.3 Attempts to map Met oxidation by Orbitrap ESI-MS/MS

To specifically map Met oxidation in tryptic Met-rhPrP^C fragments (Table 3), caused by increasing NaIO$_4$ treatment, attempts with Orbitrap ESI-MS/MS were performed (see 6.2.1). The use of Orbitrap mass spectrometry for Met oxidation mapping in tryptic fragments of rhPrP^C, was recently reported by Canello et al. (104).

The 9 Met residues are distributed on six tryptic fragments (Table 3); two of them with multiple Met residues. These are fragment 2 (Met112/129/134) and fragment 5 (Met205/206). Met- and Met(O)-peptide samples showed different retention times making semi-quantitative estimations of the oxidation state possible (see 7.4).

Fragment number	Amino acid sequence	Met number
1	TNMK	Met109
2	HMAGAAAAGAVVGGLGGYMLGSAMSRPIIHFGS	Met112/129/134
3	ENMHR	Met154
4	YPNQVYYRPMDEYSNQNNFVHDCVNITIK	Met166
5	MMER	Met205/206
6	VVEQMCITQYER	Met213

Table 3   Primary sequences of the generated Met-rhPrP^C fragments by trypsin digestion.

However, routine Orbitrap ESI-MS/MS setup (used in the Microchemistry Core Facility at MPI of Biochemistry) works in the presence of air (oxygen), which is a main source for unspecific oxidation of Met. As alternative, ESI-MS (Bruckner Daltonics) was applied since it operates exclusively under nitrogen (Fig. 16). We expected that in this experimental setup undesired oxidation would be efficiently circumvented.

In Table 4 the two analysis methods (Orbitrap ESI-MS/MS and ESI-MS) are compared, based on the oxidation level of Met (in this case of fragment 6). The analysis of Orbitrap ESI-MS/MS data was done by using the sum of the intensities of

Results

a particular fragment, with the same charge (using 1.0.9.19 MaxQuant software, Matrix Science, London, UK). In contrast, for ESI-MS analysis, the sum of the integrated peak area of a particular fragment was used for analysis. The analyzed protein molecules were not treated with any oxidant reagent in advance.

VVEQMCITQYER (Met213)	Orbitrap ESI-MS/MS	ESI-MS
Unoxidized	46.00%	100%
Oxidized **(Met)**	41.00%	0%
Dioxidized **(Met)**	13%	0%
Oxidized **(Tyr)**	2.80%	0%

Table 4 Comparison of the two instrumental approaches (Orbitrap ESI-MS/MS and ESI-MS). Met oxidation in fragment 6 was used as example.

Table 4 clearly shows that by using the Met213 fragment, Orbitrap ESI-MS/MS causes undesirable Met oxidation, of trypsinized Met-rhPrP^C. While the results gained from Orbitrap ESI-MS/MS indicated an equal proportion of unoxidized and oxidized Met in fragment 6, not even traces of oxidation of the same sample in ESI-MS were detected. Moreover, Met dioxidation as well as Tyr oxidation was observed in the Orbitrap ESI-MS/MS setup. Because of this undesirable Met oxidation, probably caused by the atmospheric airflow in the standard Orbitrap ESI-MS/MS setup, ESI-MS of Bruckner Daltonics that operates exclusively under nitrogen was used for all oxidation analyses.

Results

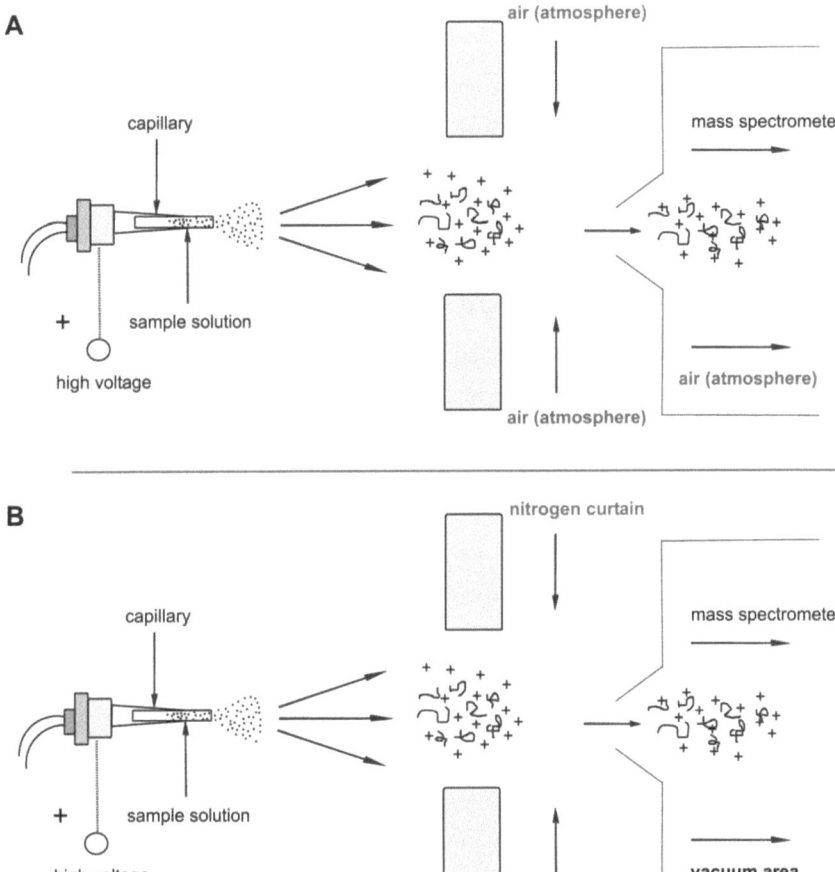

Fig. 16 Schematic representation of basic instrumental setup of Orbitrap ESI-MS/MS and ESI-MS. (A) Orbitrap ESI-MS/MS (B) ESI-MS: peptide solution is ionized, with high voltage, at the end of the capillary tube. The hereby evolving ions desolvate during the transfer into an ion trap, which captures the peptide ions using an electric field. By changing the electric field, the ions are sequentially extracted according to their mass-to-charge ratio, and are finally detected. The main difference between A and B is that in B a nitrogen curtain supports desolvation, whereas in A atmospheric air is used.

Results

7.4 Mapping Met oxidation by Bruckner Daltonics ESI-MS

The oxidation of Met-rhPrP^C was performed at 0.4 mM concentration, with 0.5, 5.0 and 25 equiv. NaIO$_4$ in 10 mM MES, pH 6.0, on ice and 12 h according to well established protocols, which proved to be well reproducible in this study (97). After tryptic digestion, the peptide mixture was separated chromatographically and was analyzed by ESI-MS (see 6.2.2 for more details).

7.4.1 Met-rhPrP^C oxidation using 5 equiv. NaIO$_4$

Expectedly, treatment with 0.5 equiv. NaIO$_4$ did not induce detectable Met-oxidation, while a 10-fold increase of oxidant (5 equiv.) yielded almost quantitative oxidation of fragment 3 (Met154) and fragment 6 (Met213). Fragment 1 (Met109), 2 (Met112/129/134) and 4 (Met166) exhibit rather low levels of modification, while no oxidation was observed for fragment 5 (Met205/206). In Fig. 17 the oxidation levels of the Met containing peptides are graphically represented, using peak area integration from mass spectra as analysis tool. The single spectra are shown in the appendix in detail (Fig. 37 - Fig. 51).

The effect of the oxidant on the protein secondary structure was further monitored by using circular dichroism. In the case of the untreated, with 0.5 equiv. and with 5 equiv. NaIO$_4$ treated Met-rhPrP^C nearly no difference in the secondary structure can be observed (Fig. 18 A, Fig. 19 A and Fig. 20 A). The secondary structure of all three show the two negative maxima at 222 and 208 nm, typical for largely α-helical proteins and are as well identical to those previously reported (76, 103). Additionally no difference was observed in their T_m values. The experiments were performed in 10 mM MES pH 6 at 37 °C (Fig. 18 B, Fig. 19 B and Fig. 20 B), as described in the Methods section (6.2.3).

Results

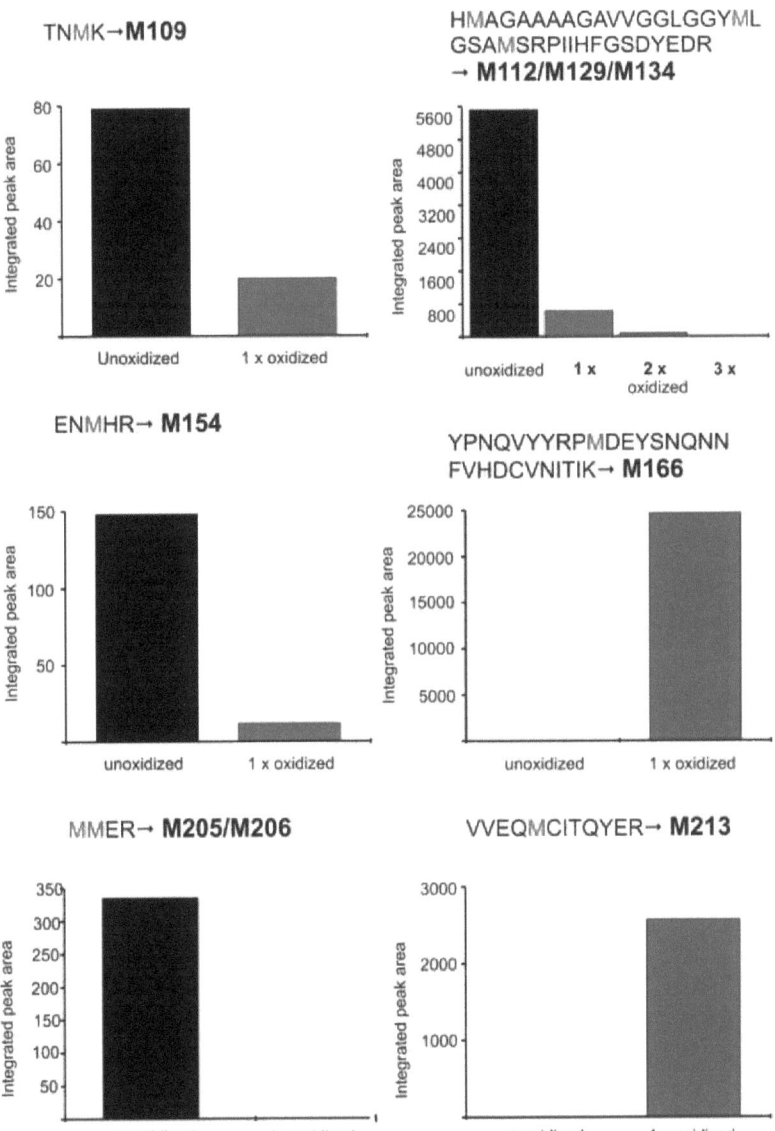

Fig. 17 **Oxidation level of the Met-rhPrP^C peptides using 5 equiv. NaIO$_4$.** The amount of oxidation was estimated by peak area integration from mass spectra. The protein was oxidized using 5 equiv. NaIO$_4$. Peptides were generated by tryptic digestion, as described in Methods (6.2.4).

Results

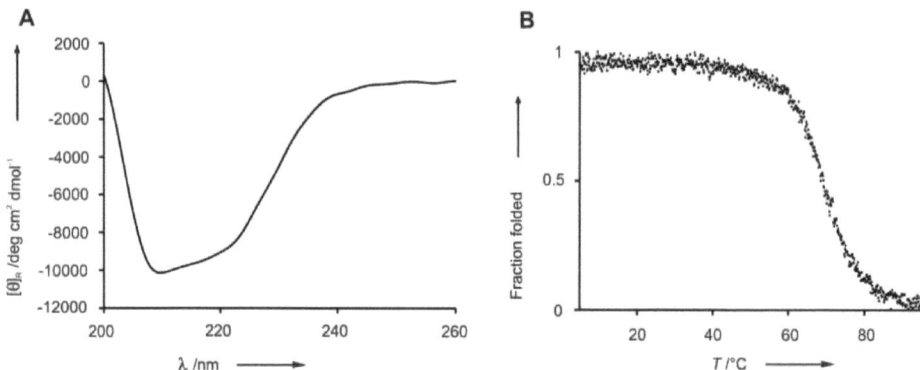

Fig. 18 Secondary structure and thermally induced unfolding profile of Met-rhPrP^C.
(A) Secondary structure of Met-rhPrP^C (c= 0.2 mg/mL), measured by far-UV CD spectroscopy.
(B) Thermally induced unfolding profile of Met-rhPrP^C (T_m = 69.4 °C). The fraction of unfolded protein was calculated from CD data monitored at 222 nm as described in Methods (6.2.3).

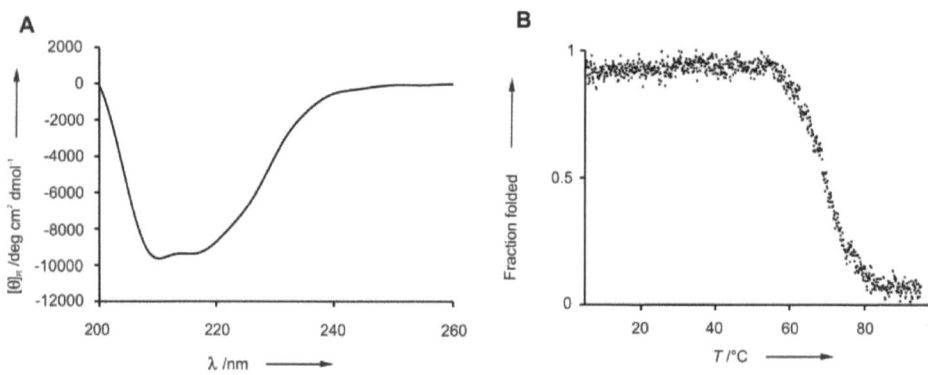

Fig. 19 Secondary structure and thermally induced unfolding profile of NaIO$_4$ with 0.5 equiv. treated Met-rhPrP^C. **(A)** Secondary structure of Met-rhPrP^C treated with 0.5 equiv.NaIO$_4$ (c= 0.2 mg/mL), measured by far-UV CD spectroscopy. **(B)** Thermally induced unfolding profile of Met-rhPrP^C treated with 0.5 equiv.NaIO$_4$ (T_m = 69.2 °C). The fraction of unfolded protein was calculated from CD data monitored at 222 nm as described in Methods (6.2.3).

Results

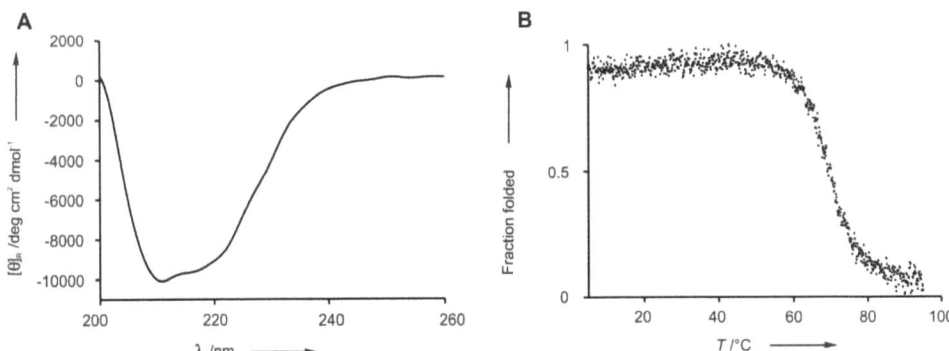

Fig. 20 Secondary structure and thermally induced unfolding profile of Met-rhPrP^C treated with 5 equiv NaIO$_4$. (A) Secondary structure of Met-rhPrP^C treated with 5 equiv.NaIO$_4$ (c= 0.2 mg/mL), measured by far-UV CD spectroscopy. (B) Thermally induced unfolding profile of Met-rhPrP^C treated with 5 equiv. NaIO$_4$ (T_m = 68.8 °C). The fraction of unfolded protein was calculated from CD data monitored at 222 nm as described in Methods (6.2.3).

7.4.2 Met-rhPrP^C oxidation using 25 equiv. NaIO$_4$ (soluble fraction)

Treatment of Met-rhPrP^C with 25 equiv. NaIO$_4$ resulted in partial precipitation. After centrifugation, both pellet and supernatant were separately analyzed. About 30% of fragment 5 (Met205/206) in the soluble fraction contained only one Met(O), whereas in the remaining 70% both Met205 and Met206 were not oxidized. The other tryptic fragments of the soluble fraction contained mainly Met(O) (Fig. 21).

The spectra of the secondary structure, in the case of the soluble part of the protein treated with 25 equiv. NaIO$_4$, showed a slight decrease in the intensity of the two negative maxima. This is an indication for a slight destabilization of the native state (Fig. 22 A). However, compared with the proteins before, as well no difference was observed in their T_m values. The experiments were performed in 10 mM MES pH 6.0 at 37 °C (Fig. 22 B).

Results

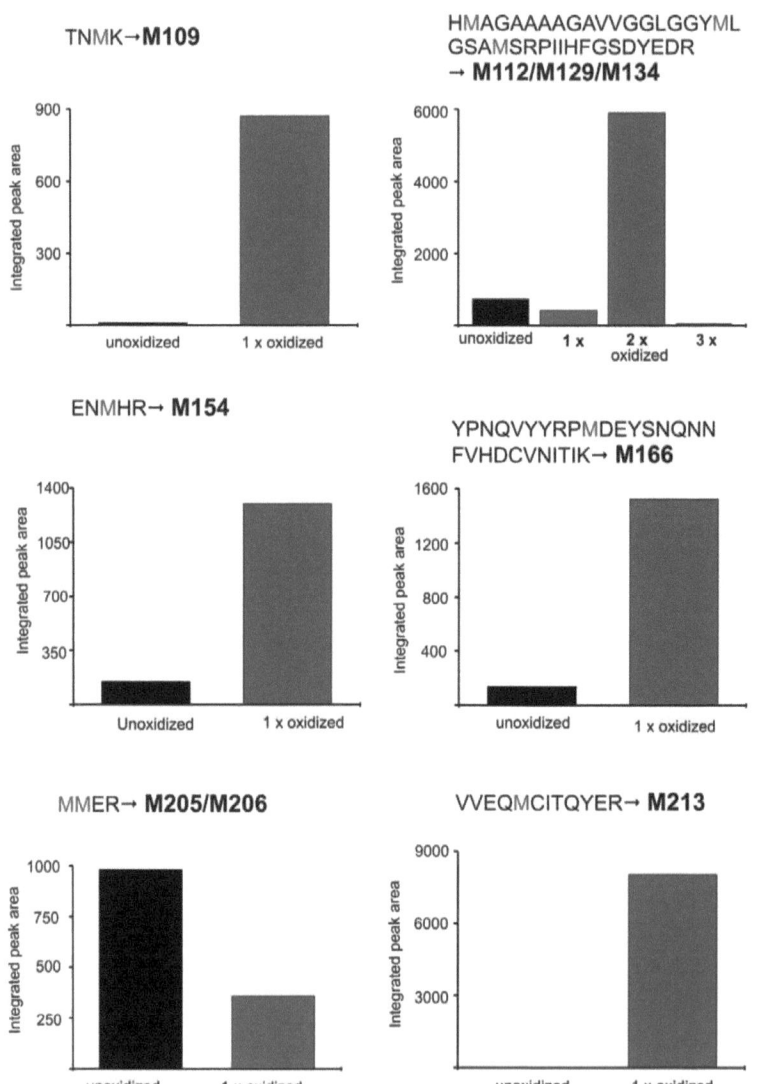

Fig. 21 Oxidation level of the Met-rhPrP^C peptides using 25 equiv. NaIO$_4$ (soluble fraction). The amount of oxidation was estimated by peak area integration from mass spectra. The protein was oxidized using 25 equiv. NaIO$_4$ (soluble fraction). Peptides were generated by tryptic digestion, as described in Methods (6.2.4).

Results

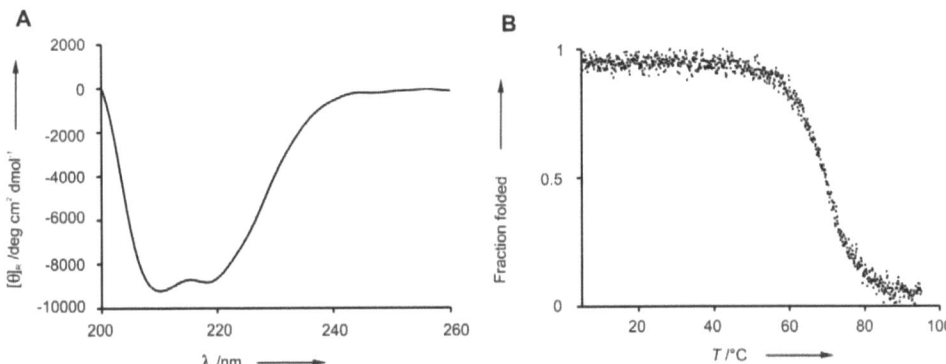

Fig. 22 Secondary structure and thermally induced unfolding profile of Met-rhPrP^C treated with 25 equiv. NaIO$_4$. (A) Secondary structure of Met-rhPrP^C treated with 5 equiv.NaIO$_4$ (c= 0.2 mg/mL), measured by far-UV CD spectroscopy. (B) Thermally induced unfolding profile of Met-rhPrP^C treated with 5 equiv. NaIO$_4$ (T_m = 69.2 °C). The fraction of unfolded protein was calculated from CD data monitored at 222 nm as described in Methods (6.2.3).

7.4.3 Met-rhPrP^C oxidized using 25 equiv. NaIO$_4$ (pellet fraction)

In the pellet (with 25 equiv. NaIO$_4$) all thioether moieties of the tryptic fragment 2 (Met112/129/134) were fully oxidized. This peptide belongs to the unstructured N-terminus. In the majority of the oxidized fragment 5 (Met205/206) from the pellet, both Met residues proved to be in the oxidized form. However, a rather high proportion of this fragment contained intact Met residues; similarly, non-oxidized Met residues were additionally found at positions 109, 154, 166 and 213. Their presence in the pellet is most probably due to the pull-down effect caused by association of soluble protein molecules to insoluble aggregates (Fig. 23).

Results

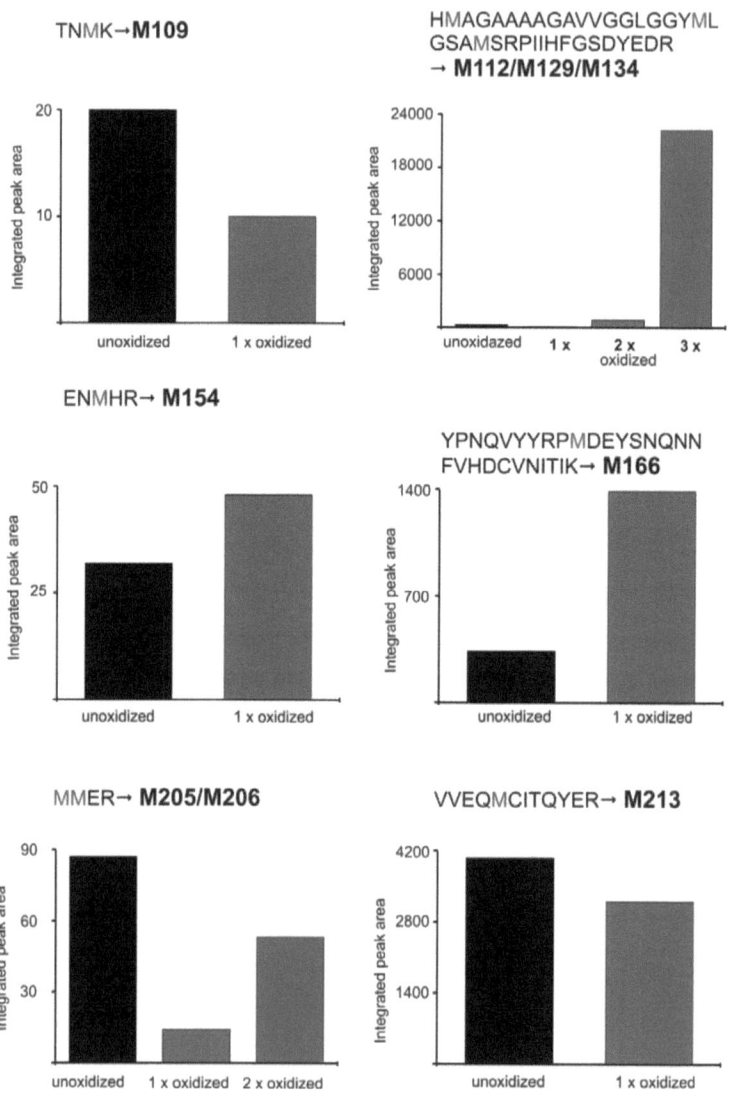

Fig. 23 Oxidation level of the Met-rhPrP^C peptides using 25 equiv. NaIO$_4$ (pellet fraction). The amount of oxidation was estimated by peak area integration from mass spectra. The protein was oxidized using 25 equiv. NaIO$_4$ (pellet fraction). Peptides were generated by tryptic digestion, as described in Methods (6.2.4).

Results

7.4.4 Overall picture of NaIO₄ induced Met oxidation

The overall pattern of the NaIO$_4$ induced oxidation in all analyzed peptides is presented in Fig. 24. These findings confirm that the buried Met205/206 residues are less accessible to oxidants even when increasing the concentration to 25 equiv. This fact agrees with previous reports from other laboratories (102, 103). With this optimized analytical method rather confident quantification of the extent of oxidation of individual Met residues in rhPrP^C by NaIO$_4$ were obtained.

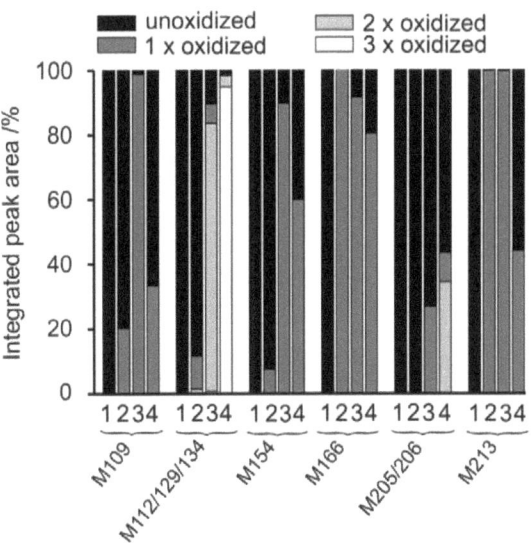

Fig. 24 Graphical representation of the overall oxidation results in Met containing peptides. The extent of Met oxidation was evaluated from the integrated peak areas: (1) without NaIO$_4$, (2) with 5 equiv. NaIO$_4$, (3) with 25 equiv. NaIO$_4$ (soluble fraction), and (4) with 25 equiv. NaIO$_4$ (pellet fraction).

7.5 Attempts for Met(O) and Met(O₂) incorporation into rhPrP^C

In order to check the hypothesis that oxidation of the buried Met residues could be the critical event *in vivo* for the α → β structural transition in prion protein, we attempted first to express rhPrP^C with all Met residues fully replaced by Met(O) or

Results

even Met(O$_2$). Unfortunately, the incorporation of Met(O) was less successful, probably mainly due to the intracellular activity of Msr (see 3.6.4) and reducing activity of the bacterial cytosol. Therefore, in the mass spectra not only the wild type protein (Met-rhPrP^C) and the fully exchanged Met(O)-rhPrP^C were present, but also all different species between them as well as all possible adducts. In the other words, the generation of high-level labeled homogeneous Met(O)-protein sample was not possible. The best result of all these attempts is presented in Fig. 25. On the other hand, all attempts to incorporate Met(O$_2$) failed.

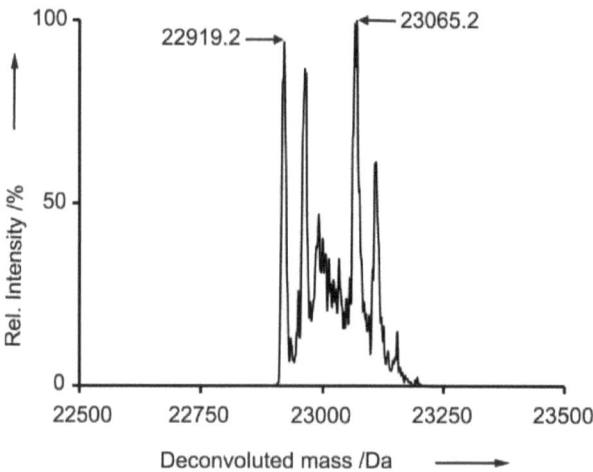

Fig. 25 Deconvoluted mass spectrum of protein sample containing a portion of fully labeled Met(O)-rhPrP^C. The mass of 22919.2 Da corresponds to Met-rhPrP^C, while the peak with 23065.2 Da represents the complete exchange of Met to Met(O) in the protein. Other accompanying peaks between Met-rhPrP^C and Met(O)-rhPrP^C represent all possible labeling levels as well as adducts generated due to the binding of Na$^+$-ions from buffer.

Since it was not possible to incorporate neither Met(O) nor Met(O$_2$) in a way to have a suitable homogeneous protein samples for analyses, an alternative method that is capable to mimic chemically Met oxidation and reduction, had to be found.

7.6 Expression and isolation of Nle-rhPrP^C and Mox-rhPrP^C

All Met residues in rhPrP^C were substituted by chemically stable and translational active analogs, which were capable to mimic the reduced and oxidized protein state. As best candidates the isosteric Met analogs norleucine (Nle) and methoxinine (Mox), respectively, were selected (Fig. 26).

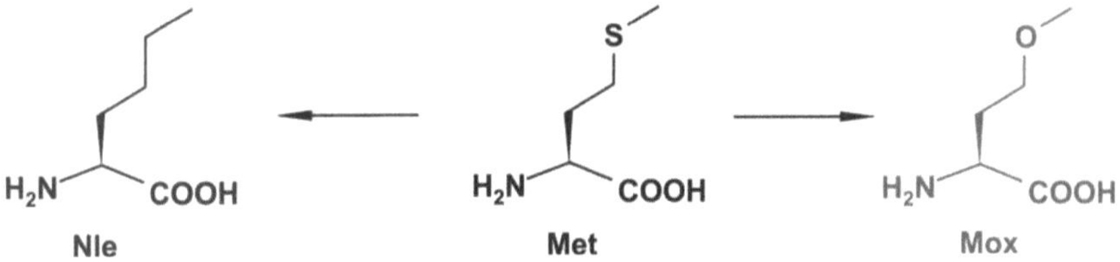

Fig. 26 Methionine (Met) and its isosteric analogs norleucine (Nle) and methoxinine (Mox).

First experiments to achieve high-level incorporation of the Mox analog, based on the supplementation incorporation (SPI) method (86), was not successful (Fig. 27). The incorporation of Mox is much more difficult than the incorporation of Nle, although in both cases heterogeneous protein samples were often formed as shown in Fig. 27. However, the incorporation of both analogs can be subsequently improved after optimization work (for detail description see Methods). This study is the first report, where noncanonical amino acid Mox was successfully incorporated into proteins (Fig. 28 C). As shown in Fig. 29 the expression efficiency of Met-rhPrP^C and its variants was approximately the same. Furthermore, the yields of the purified proteins were sufficient.

Results

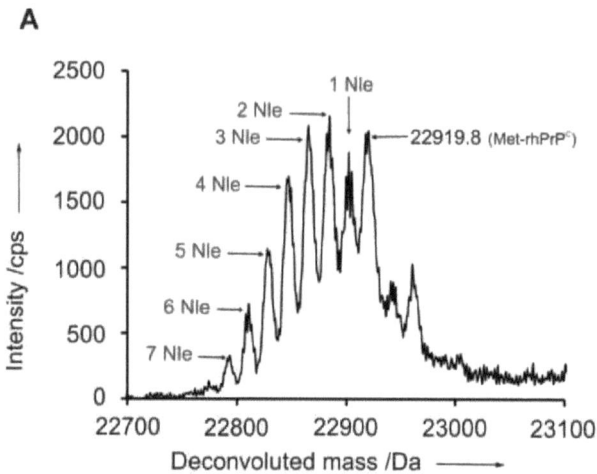

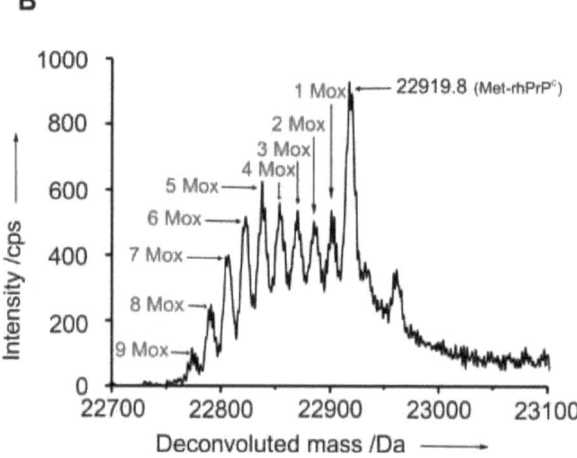

Fig. 27 Deconvoluted mass spectra of Nle-rhPrP^C and Mox-rhPrP^C – heterogeneous masses.
(A) Mass profile for incomplete incorporation of Nle in rhPrP^C. The peak at 22919.8 Da corresponds to Met-rhPrP^C, whereas the other peaks illustrate the different incorporation levels (from 1 to 7 Met → Nle exchanges). **(B)** Mass profile for incomplete incorporation of Mox in rhPrP^C. The peak at 22919.8 Da corresponds to Met-rhPrP^C, whereas the other peaks illustrate the different incorporation levels (from 1 to 9 Met → Mox exchanges).

Results

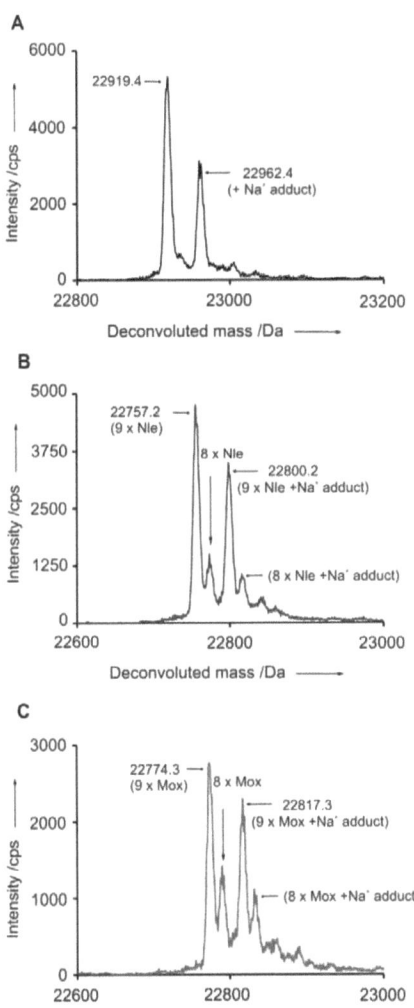

Fig. 28 Deconvoluted ESI-MS spectra of Met-rhPrP^C – high level incorporation (A), Nle-rhPrP^C (B) and Mox-rhPrP^C (C). (A) The predominant peak corresponds to the mass of Met-rhPrP^C. The accompanying peak is most probably unspecifically Na$^+$-adducts from buffer. (B) High level incorporation of Nle into rhPrP^C. Full labeled mass species (9 x Nle) dominates the chromatogram. The mass of Nle-rhPrP^C corresponding to 8 exchanged Met residues is also detectable. All other species are protein-Na$^+$-adducts. (C) High level of Mox → Met substitution in rhPrP^C is characterized by stronger presence of mass species with 8 exchanged Met residues.

Results

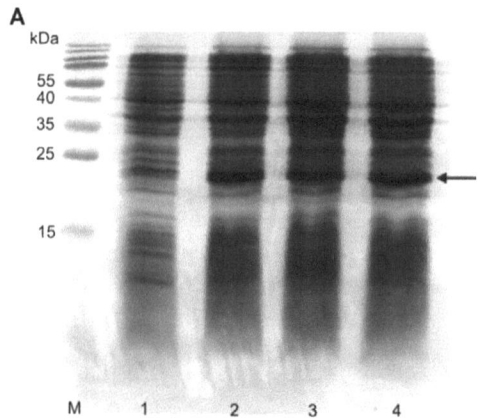

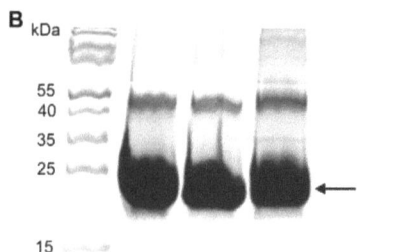

Fig. 29 Coomassie stained SDS-PAGE gels of Met-rhPrP^C and analogs. (A) SDS-PAGE of expression profiles. M: marker; 1: non induced; 2: Met-rhPrP^C; 3: Nle-rhPrP^C; 4: Mox-rhPrP^C. The arrows point at the protein bands. **(B)** SDS-PAGE of purified samples. M: marker; 1: non induced; 2: Met-rhPrP^C; 3: Nle-rhPrP^C; 4: Mox-rhPrP^C. The arrows point at the correct protein bands.

The mass difference between Met-rhPrP^C and its two variants suffices in both cases for analysis by ESI-MS. The mass found for Met-rhPrP^C was 22919.2 Da (Fig. 28 A), which is in excellent agreement with the theoretically expected value. As observed previously with other systems (89), the substitution of the N-terminal Met in rhPrP^C with Nle and Mox does not prevent their excision. The Met → Nle substitution lowers the molecular mass of the protein by 18 Da per Met residue. The expected and found mass for Nle-rhPrP^C was 22757.2 Da (Fig. 28 B). From the intensities of

Results

the mass ion signals, we could estimate semi-quantitatively a high Nle incorporation (~ 97%). Similarly, the molecular weight change from Met to Mox is 16 Da, which corresponds to a difference of 144 Da for the whole recombinant protein. Indeed, we found the expected mass of 22774.3 Da for Mox-rhPrP^C (Fig. 28 C). However, the level of substitution was somewhat lower (~ 85 %) than for the Nle-variant. Further analytical characterization was achieved by tryptic peptide fragment sequencing.

7.7 CD spectroscopy of Nle-rhPrP^C and Mox-rhPrP^C

The far-UV CD spectrum of Met-rhPrP^C at 37 °C in 10 mM MES at pH 6.0 is identical to those previously reported (76, 103) and has the two negative maxima at 222 and 208 nm, typical for largely α-helical proteins. Compared to the parent Met protein, the spectrum of the Nle-variant (Fig. 30) shows increased intensities of the two maxima by about 10% indicating further stabilization of the native state. Conversely, for the Mox-rhPrP^C, significant changes in the dichroic properties were observed. The negative maximum is shifted to 215 nm and the overall intensity is markedly reduced. Obviously, the global Met → Mox replacement in the prion protein leads to conversion of the prevalent α-helical structure to β-type conformations.

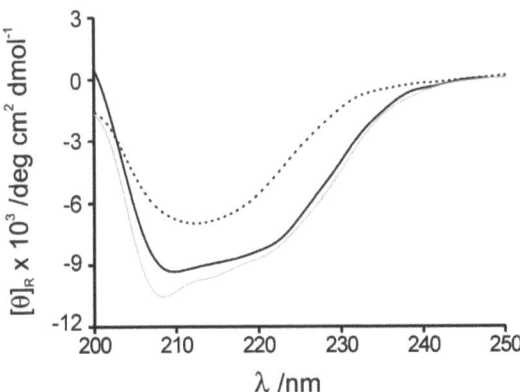

Fig. 30 CD spectra of Met-rhPrP^C and its Nle- and Mox-variants at 37 °C and 0.2 mg/mL in 10 mM MES pH 6.0. Met-PrP^C: ——; Nle-PrP^C: ——; Mox-PrP^C: — — — . Experiments were performed as described in Materials.

Results

7.8 Monitoring secondary structure change upon heating

Moreover we monitored variations of spectral shape and intensities (i.e. protein secondary changes), in the far-UV CD region, upon heating. The dichroic spectra of Met-rhPrP^C at 4 °C and 37 °C are nearly identical, in spite of the substantial temperature differences. This indicates a relatively compact folded state. At a temperature range around 60 °C transition to a denatured state occurs (Fig. 31 A). The complete denaturation of Met-rhPrP^C secondary structure is observed at 95 °C.

The comparison of the Nle-rhPrP^C CD (Fig. 31 B) curve shape, with those of the parent protein, showed similar signal ratio between the intensities of the two minima. Both, Met-rhPrP^C and Nle-rhPrP^C, have a compact set of dichroic curves in the native state. This is characterized by a rather sharp transition to the denatured state. However, in Nle-rhPrP^C the intensities of the minima were increased (~10%), which indicates further stabilization in the native state of this protein. This remarkable feature becomes obvious upon heating of Nle-rhPrP^C, whereby its secondary structure content starts to change significantly only after 60 °C. It is also noteworthy that the dichroic intensity band around 208-210 nm in Nle-rhPrP^C variant is more distinctive than this of the parent protein. Conversely, in the CD spectrum of Mox-rhPrP^C (Fig. 31 C) radical differences compared to the dichroic profiles of Met-rhPrP^C and Nle-rhPrP^C were observed. These are characterized by a substantial reduction in the overall Mox-rhPrP^C dichroic intensity of ~ 40% and by the emergence of novel negative maxima at 215 nm. Obviously, the global Met → Mox replacements in the whole prion sequence are responsible for the high β-sheet content in Mox-rhPrP^C.

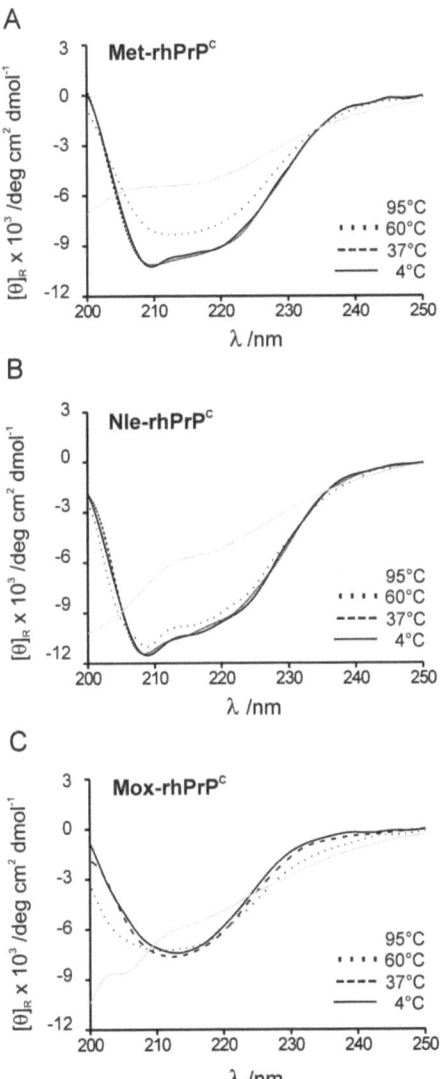

Fig. 31 Secondary structure changes as a function of temperature increase, in Met-rhPrP^C and related variants, measured by far-UV CD spectroscopy. (A) Met-rhPrP^C; (B) Nle-rhPrP^C; (C) Mox-rhPrP^C. Experiments were performed as described in Methods.

Results

7.9 Melting curves of Met-rhPrP^C and its variants

The temperature-induced unfolding of Met-, Nle- and Mox-rhPrP^C was monitored by recording the dichroic intensities at 222 nm (Fig. 32). Since thermal unfolding of these proteins leads to irreversible denaturation, thermodynamic parameters could not be derived. As already suggested by the increased dichroic intensities at 37 °C the enhanced stability of Nle-rhPrP^C is well reflected by the higher melting point (T_m = 74.4 ± 3.4 °C) compared to Met-rhPrP^C (T_m = 65.2 ± 4.2 °C) (Fig. 32 A). An enhanced thermal stability induced by the Met → Nle replacement has previously been observed for the α-helical annexin A5 (105) and has been attributed to the increased hydrophobicity of the protein core by the buried Nle residues. Similarly, the buried norleucines 205/206 should strongly contribute to the markedly enhanced stability of the Nle-rhPrP^C variant.

For the thermal unfolding of Mox-rhPrP^C a gradual, non-cooperative melting between 42 °C and 95 °C was observed (Fig. 32 B). Such unfolding patterns are typical for proteins, which either are very flexible and have a partially unfolded ground state or contain heterogeneous populations of folded states. We attribute the different unfolding behavior of Mox-rhPrP^C as compared to Met- or Nle-rhPrP^C to its flexibility and the mixed populations of prevalently β-sheet structure of this variant.

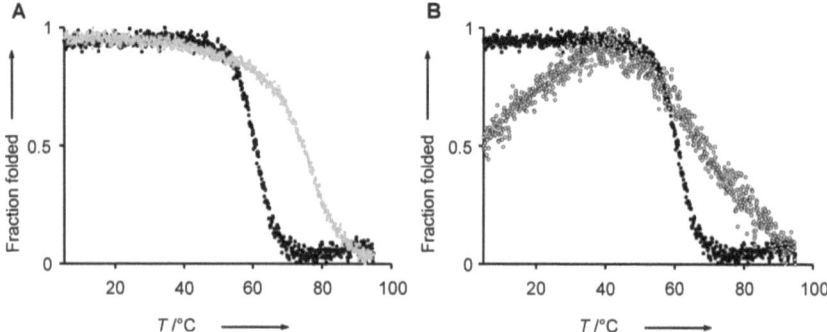

Fig. 32 Thermal denaturation monitored by the changes of dichroic intensities at 222 nm in function of temperature. **(A)** Comparison between the melting curve of Met-rhPrP^C (•••black) and Nle-rhPrP^C (••• grey). **(B)** Comparison between the melting curve of Met-rhPrP^C (•••black) and Mox-rhPrP^C (○○○ white).

An additional striking feature of the temperature induced denaturation experiment with the Mox-variant is the maximum stability in the temperature range between 35 °C and 45 °C, while below and above these temperatures protein denaturation takes place (Fig. 32 B). A decrease in protein stability induced by lower temperatures is known as cold denaturation (106). In natural proteins cold denaturation usually occurs below the freezing point of water and was observed for proteins with hydrophilic amino acid residues in the core structure (107). Therefore, the cold-denaturation-like melting of Mox-rhPrP^C is most probably caused by the introduction of hydrophilicity in the globular domain of rhPrP^C with the methoxinine 205/206 residues. A similar observation was reported recently for a variant of the ribonuclease inhibitor bastar containing a hydrophilic tryptophan analog in the protein interior (27).

7.10 Pro- and anti-aggregation prion protein variants

The *in vitro* aggregation tendency of Met-rhPrP^C and its variants was monitored using the well established *de novo* SDS-dependent *in vitro* aggregation assay (87, 101). The aggregation behavior of fluorescently labeled rhPrP^C was observed by using confocal single molecule analysis techniques; such as fluorescence cross-correlation spectroscopy analysis (cross-correlation FCS) as well as scanning for intensely fluorescent targets (SIFT) (91, 98). The samples containing 0.2% SDS act as reference, since in 0.2% SDS no aggregation should take place (data not shown). The aggregation onset was, when the SDS concentration in the samples was diluted to 0.0125%. In Fig. 33 A, the aggregation capacity of Nle-rhPrP^C is significantly reduced compared to that of Met-rhPrP^C, under the identical experimental conditions. This strong anti-aggregation effect directly correlates with an enhanced stability of the spatial structure and a high α-helical content of Nle-rhPrP^C (*vide infra*). In contrast, the Met replacement by Mox in rhPrP^C caused a strong pro-aggregatory effect. As expected, in the aggregation assay under oxidative conditions both Nle- and Mox-variants were relatively insensitive to the different NaIO$_4$ concentrations (Fig. 33 B). The observed increase in the aggregation tendency, at higher oxidant concentrations (over 2 mM), has to be attributed to the oxidative degradation of other

Results

sensitive residues, particularly Trp and Tyr (103). Thus, Nle-rhPrP^C and Mox-rhPrP^C proved to be suitable tools to investigate the effect of Met oxidation.

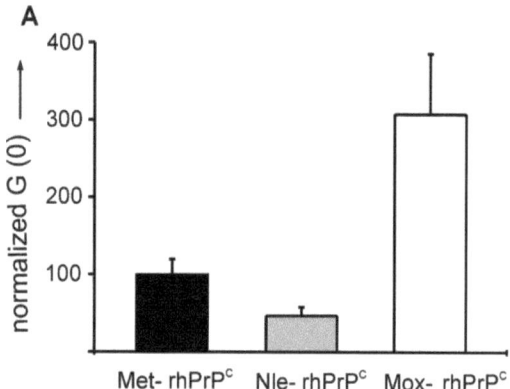

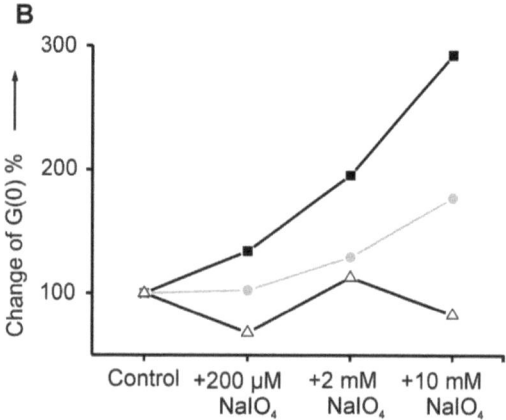

Fig. 33 *In vitro* aggregation of Met-rhPrP^C compared to Nle-rhPrP^C and Mox-rhPrP^C. (**A**) *In vitro* aggregation assay performed in the absence of oxidants. In the presence of 0.2% SDS no aggregation is observed in all three samples (data not shown). Normalizes G(0) of Met-rhPrP^C is set to 100%. (**B**) Aggregation tendency as a function of NaIO$_4$ concentration. The intrinsic aggregation tendency of all three samples in the absence of NaIO$_4$ is arbitrarily set to 100%. Data points represent the mean of 6 different measurements. Met-rhPrP^C: ■ squares; Nle-rhPrP^C: ● circles; Mox-rhPrP^C: △ triangles.

Results

7.11 Far-UV CD spectroscopy of the designed Nle and Mox peptide

From peptide studies Met is well known to stabilize and efficiently induce α helical conformations (108, 109). Similarly, Nle exhibits an even higher intrinsic preference for α-helical states but lower preferences for β-sheet conformations than Met residues (110). In the absence of experimental data on structural propensities of Mox residues, which conceivably could prefer β-sheet over α-helical conformation because of their hydrophilicity, we have addressed this question with the suitable model peptide Ac-YLKAMLEAMAKLMAKLMA-NH$_2$ of Dado and Gellman which showed an α → β transition upon Met-oxidation (66). The Nle-peptide exhibits the typical α-helical CD profile with a very high content of ordered structure (> 80% α-helix) (Fig. 34). Conversely, the related Mox-peptide shows a dramatically decreased α-helical content (~ 40%) with a 6-fold increase in random coil, but also a significant percentage of β-type structure (20 %) (Table 5). Taken together, one can conclude that Mox represents a good mimic for Met(O).

Nle-peptide: Ac-YLKA**Nle**LEA**Nle**AKL**Nle**AKL**Nle**A-NH$_2$

Mox-peptide: Ac-YLKA**Mox**LEA**Mox**AKL**Mox**AKL**Mox**A-NH$_2$

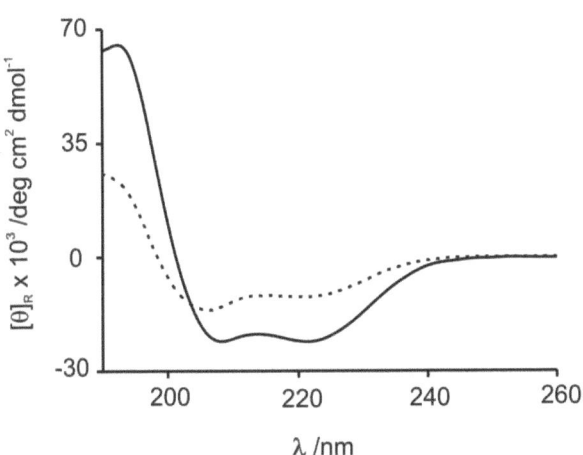

Fig. 34 Sequence and secondary structure of two variants, Nle (—) and Mox (---), of the 18 amino acid model peptide. The measurement was done by far-UV CD spectroscopy with a concentration of 100 μm at 20 °C.

Results

190-260 nm	Nle [%]	Mox [%]
Helix	84.3	38.1
Antiparallel	0.1	3.8
Parallel	1.6	6.8
Beta-Turn	8.7	19.2
Rndm. Coil	5.3	32.1

Table 5 Calculated composition of the different secondary structures of the two peptide analogues, Nle (blue) and Mox (red), of the Dado-Gellmann model peptide. Computed with the CONTIN software (Reference SMP56)/CDPro Analysis.

8 Discussion

8.1 Met oxidation as a possible origin of prion protein structural conversion

The formation of PrP^{Sc} from PrP^{C} is a post-translational process that is believed to follow an autocatalytic mechanism (82). The conversion involves a conformational change in which the α-helical content of the protein decreases while the amount of β-sheet dramatically increases (85). A chemical modification of particular residues might trigger this α → β transition. Until now, no candidate for chemical modification has been unambiguously identified. One suspicious candidate could be Met, since the oxidation of its side chain changes the conformational preferences in model peptides (66). Additionally, Met is readily oxidized by most reactive oxygen species. Such chemical modification (induced by ROS) of all Met residues (especially buried ones) might represent an initial event that leads to intramolecular α → β structural conversion and subsequent fatal PrP^{C} → PrP^{Sc} transition. Although there is increasing evidence for an important role of PrP^{C} in oxidative stress (83), a direct correlation between PrP^{C} oxidation and its conversion to a β-sheet rich form, so far, was not clearly established. Therefore, the aim of this work was to shed light on this hypothesis.

8.2 Recombinant hPrP^{C} as model for structural conversion

In nature, the cellular prion protein is an N-linked glycoprotein normally bound to the neuronal cell membrane by a GPI anchor (78). However, for structural studies recombinant PrP^{C} is favored, most notably because of the difficulty in isolating and purifying the necessary amounts of PrP^{C} from tissue. Moreover the combination of circular dichroism, ^{1}H-NMR spectroscopy (111), antibody-binding studies (112) as well as molecular dynamics simulations (113), indicate that rPrP^{C} (unglycosylated and without GPI anchor) and the natural glycoprotein (glycosylated and with GPI anchor) share similar structural characteristics. Furthermore, unglycosylated isoforms of PrP^{C} exist as well *in vivo* and their conversion to PrP^{Sc} is confirmed (114). Recently a number of experimental and theoretical studies investigated the possible role of glycosylation and membrane anchoring on PrP^c structure, as observed in the work from Ollesch and associates (115), which found that membrane anchoring of prion protein at high concentration profoundly alters its secondary structure. However, other studies such as from DeMarco and Deggett (113) reported that glycosylation and attachment of PrP^C to the membrane surface via a GPI anchor, does not significantly change the structure or dynamics of PrP^C. Independent of this unresolved issue, it is unlikely that the state of the protein in solution or in the membrane-anchored form has an influence on possible structural perturbation caused by the oxidation of Met residues localized in the folded, exposed part of the protein. Accordingly, all experiments were performed with $rhPrP^C$.

8.3 Protein damage caused by Met oxidation

Diverse human disorders, including several neurodegenerative diseases, systematic amyloidoses and age-related diseases, arise from the misfolding and aggregation of an underlying protein (53). The role of oxidative damage caused by Met oxidation in these processes is scarcely appreciated in the contemporary literature. However, since Met with its high oxidation propensity is one of the major targets of ROS, the consequences of its oxidative modifications cannot easily be neglected. The oxidation of surface-exposed Met residues was suggested to represent an endogenous antioxidant defense (scavenger function) which protects proteins against the oxidation by ROS (116). The degree of oxidized Met residues in cells is controlled by the balance of the production of ROS and the reduction of Met(O) back to Met by Mrs (Fig. 6) (57). Therefore, diseases related to protein misfolding might be a logical consequence of an imbalance between cellular oxidation and reduction reactions and/or a loss of other protective mechanisms.

The accumulation of oxidized proteins, especially those containing oxidized Met residues is also a hallmark of aging (117). Notably, the performance of the Msr enzymatic system (Fig. 6) determines aging, stress resistance and lifespan of bacteria, yeast, insects and mammals (68). For example, in transgenic *Drosophila* overexpression of MsrA extended the mean life span up to 70%, whereas mice without this gene had on average about 40% shorter life spans (57, 58).

Discussion

8.4 Oxidation of Met residues in rhPrP^C and structural conversion

8.4.1 Met residues in the prion protein

The Met residues in rhPrP^C are mainly surface-exposed and located in the structured globular part (125-231) as follows (Fig. 7): Met129 in the β-strand I, Met134 in the loop between β-strand 1 and α-helix I, Met154 in α-helix I, and Met166 in β-strand II. Only Met109 and Met112 are located in the unstructured N-terminus (23-124). However, Met205 and Met206 in α-helix III are buried, whereas Met213, as well located in α-helix III, is only partially surface-exposed.

The left-handed β-helical model of the human PrP89-146 by Langedijk et al. (118) places particular emphasis on the role of Met109 and Met129 in providing stability for the formation of a stable left-handed β-helical structure. Interestingly, Met129 is as well involved in prion protein polymorphism, Met or Val at codon 129 (Met129Val) and Glu or Lys at codon 219 (Glu219Lys) (114). The common human Met–Val polymorphism at position 129 has a profound influence on prion pathogenesis. For instance, heterozygosity for Met and Val at position 129 is highly protective against sporadic and acquired prion diseases in humans (119). On the contrary, nvCJD occurs exclusively in Met129 homozygotes (120).

Position 205 is invariantly occupied by hydrophobic residues in prion proteins from different species and is well conserved in all mammalian prion proteins. Met205 is part of the hydrophobic face of helix III and is involved in a network of interactions that stabilize the packing of this structural motif. Its replacement by hydrophilic Ser or Arg prevents folding *in vivo* of the mutant protein (121), a fact which was confirmed by molecular dynamic simulations (122). A similar structural role can possibly be assigned to the vicinal Met206 residue. Indeed, by using molecular dynamics simulations, Colombo and coworkers suggested recently that oxidation of helix III Met residues (Met206, Met213) might be the switch triggering the initial α-fold destabilization that is required for productive pathogenic conversion of prion protein (123).

Discussion

8.4.2 Structural consequences of Met oxidation in the prion protein

The aforementioned facts led us to propose the following hypothesis: The oxidation of PrP^C, especially of Met205/206 in the hydrophobic core dramatically changes the intrinsic local conformational preferences and facilitates global α → β structural conversion in prion protein, which consequently leads to aggregation and disease.

The distribution pattern of Met residues in the prion protein sequence certainly contributes to the local organization of the secondary structure elements (α-helices and β-sheets). It was shown, by Dado and Gellman, for model peptides that Met favors α-helices whereas oxidized Met induces β-sheets (66). In order to assess whether Met oxidation similarly triggers α → β conversion in prion protein, we specifically oxidized (97) the Met residues in rhPrP^{C}23-231 with increasing amounts of sodium periodate (NaIO$_4$).

8.4.3 Prion protein aggregation upon periodate oxidation

Protein aggregation is a central aspect, regarding neurodegenerative diseases in general and prion disease in particular (124). Therefore we wanted to examine, if specific oxidation of the Met residues in rhPrP^C with periodate (97), induces prion protein aggregation. To achieve this, a *de novo* aggregation method developed by the group of D. Riesner (101, 125) was employed. In recent years, FCS and SIFT have been recognized as methods that allow highly sensitive analysis of protein aggregation in neurodegenerative diseases such as prion diseases at the molecular level.

The oxidation experiments showed not only an increase in aggregation tendency of rhPrP^C with increasing periodate concentration (Fig. 15), but also a higher proportion of oxidized Met residues at elevated NaIO$_4$ equiv. was observed (see 7.4.4). While at lower oxidant concentration, e.g., 5 equiv. NaIO$_4$, mainly the surface-exposed Met residues were oxidized to Met(O). With 25 equiv. NaIO$_4$, in the soluble fraction, even the buried Met205/206 residues were at least partially oxidized as was assessed by the ESI-MS analysis of tryptic digests (Fig. 35).

Discussion

In the 25 equiv. NaIO$_4$ pellet fraction the Met205/206 tryptic peptide was found to contain one Met(O), as well as two Met(O)s (Fig. 35). Since we found the heavy oxidation of Met205/206 predominantly in the pellet fraction we postulate that the oxidation of these buried Met residues leads to precipitation of rhPrP^C. The concomitant relatively high proportion of unoxidized Met205/206 in the the same fraction can be ascribed to the pull down effect of 'healthy' molecules during precipitation.

From these results, it is conceivable that PrP^C is tolerant toward individual Met-oxidations substitution, especially at the surface. However, it might not be plastic enough, to tolerate total oxidation of all individual Met residues, without devastating consequences. Thus, we believe that once such profound chemical modification (induced by ROS) occurs at all Met residues (especially buried ones) the intramolecular α → β structural conversion inevitably takes place. Therefore, the conversion of moderately hydrophobic Met to hydrophilic Met(O) causes the secondary structure rearrangement in PrP^C.

Discussion

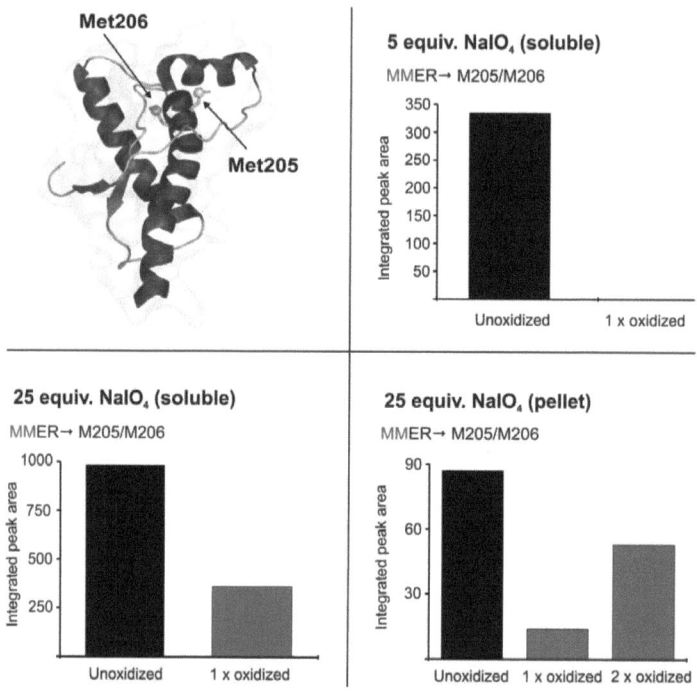

Fig. 35 **Comparison of the Met oxidation level in the Met205/206 peptide, using different NaIO$_4$ equivalents.** On the upper left 3D-structure of human PrP^C(125-231) (PDB accession No. 1QM0) with buried Met-residues (205 and 206). Note that sample aggregation takes place only when 2 x oxidized **MMER** peptide species are detectable.

The oxidation of these critical residues provokes precipitation, probably because of enhanced aggregation. However, a quantitative correlation between oxidation state of the Met residues and aggregation propensity of rhPrP^C is prevented by the experimentally different conditions required for the aggregation assay, i.e. the very low protein concentration and thus strongly reduced reaction rates that require higher oxidant excesses, and the assessment of periodate Met oxidation by ESI-MS. Nevertheless, by increasing the NaIO$_4$ amounts a significantly enhanced aggregation propensity was observed (Fig. 15). A contribution of oxidative degradation of sensitive residues such as His, Trp and Tyr cannot be excluded in the performed *de novo* aggregation assay (102, 103).

Discussion

8.5 Chemical model for α → β conversion in rhPrP^C

After we had observed the intriguing aggregation effect that periodate oxidation of Met residues produces on rhPrP^C, we sought to delineate how we could induce the intramolecular α → β structural conversion of rhPrP^C in the frame of a well-defined model system. The classical site directed mutagenesis is less suitable for that purpose since observed folding properties cannot unambiguously be assigned to the different shape of the new side chain or to its chemistry. Attempts to study the role of Met residues in proteins by classical site directed mutagenesis in the frame of 20 canonical amino acids are often controversial. For example in a Met205Arg mutant (122) it is difficult to assess, whether the observed folding properties are due to the differences in the shape of the side chain or to the chemical differences of the atoms involved.

In order to create a model system that allows a direct correlation between Met oxidation and aggregation propensity, we first aimed at the quantitative substitution of Met by Met(O) or Met(O$_2$) in the prion protein. Unfortunately, due to the intracellular catalytic activity of Msr, as well as general reducing features of cytosol, our attempts to replace all Met residues in rhPrP^C with Met(O) or Met(O$_2$), respectively, by the SPI method failed. Therefore, we selected the alternative approach of replacing the Met residues with synthetic Met analogs. For our study we replaced the Met residues by the significantly more hydrophilic Mox to mimic the Met(O)-rhPrP^C form and additionally combined it with the expression of the oxidation-insensitive Nle-variant (Fig. 36). The residue specific replacement of these isosteric Met analogs in the prion protein proved to be a suitable tool to study the effects of extreme hydropathy changes (hydrophilic Mox mimicking Met(O) vs. hydrophobic Nle mimicking Met) on protein aggregation. Both Mox and Nle have nearly identical chain lengths and are resistant to chemical oxidation and reduction of their side chains. By introducing such modest alterations into the studied protein structures, possible steric effects should be minimized. Thus, any structural differences in the resulting Mox and Nle protein variants reflect only the possible conformational preferences provoked by the hydropathy change.

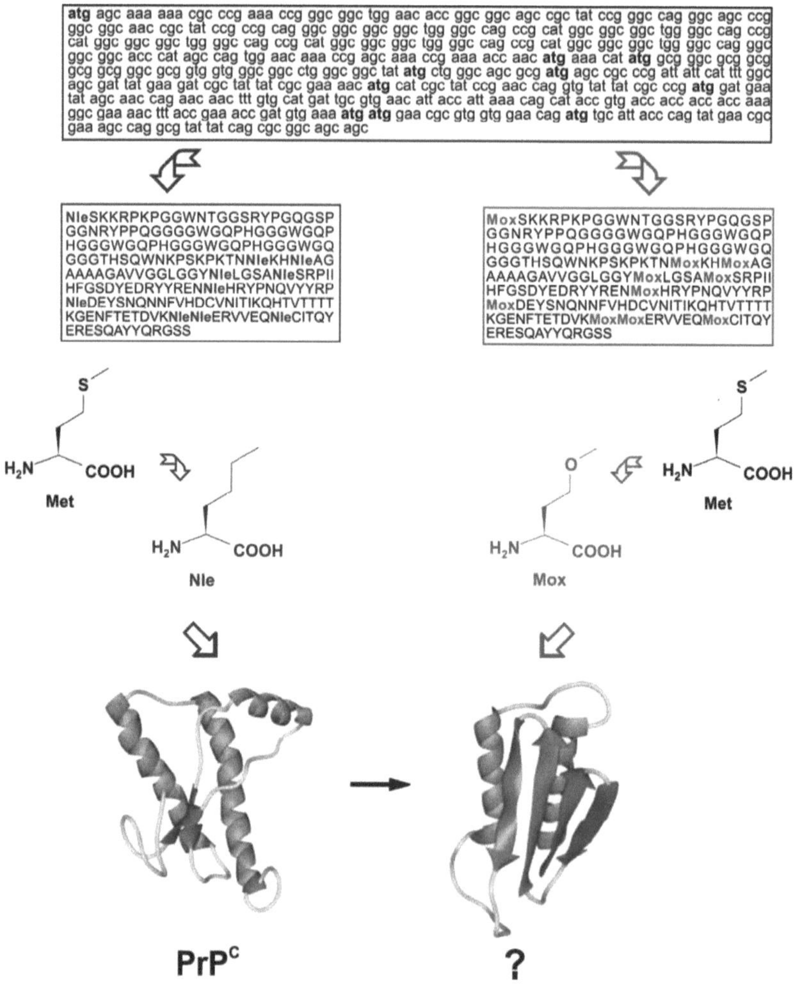
atg agc aaa aaa cgc ccg aaa ccg ggc ggc tgg aac acc ggc ggc agc cgc tat ccg ggc cag ggc agc ccg
ggc ggc aac cgc tat ccg ccg cag ggc ggc ggc ggc tgg ggc cag ccg cat ggc ggc ggc tgg ggc cag ccg
cat ggc ggc ggc tgg ggc cag ccg cat ggc ggc ggc tgg ggc cag ccg cat ggc ggc ggc tgg ggc cag ggc
ggc ggc acc cat agc cag tgg aac aaa ccg agc aaa ccg aaa acc aac atg aaa cat atg gcg ggc gcg gcg
gcg gcg ggc gcg gtg gtg ggc ggc ctg ggc ggc tat atg ctg ggc agc gcg atg agc cgc ccg att att cat ttt ggc
agc gat tat gaa gat cgc tat tat cgc gaa aac atg cat cgc tat ccg aac cag gtg tat tat cgc ccg atg gat gaa
tat agc aac cag aac aac ttt gtg cat gat tgc gtg aac att acc att aaa cag cat acc gtg acc acc acc acc aaa
ggc gaa aac ttt acc gaa acc gat gtg aaa atg atg gaa cgc gtg gtg gaa cag atg tgc att acc cag tat gaa cgc
gaa agc cag gcg tat tat cag cgc ggc agc agc
NleSKKRPKPGGWNTGGSRYPGQGSP
GGNRYPPQGGGGWGQPHGGGWGQP
HGGGWGQPHGGGWGQPHGGGWGQ
GGGTHSQWNKPSKPKTNNleKHNleAG
AAAAGAVVGGLGGYNleLGSANleSRPII
HFGSDYEDRYYRENNleHRYPNQVYYRP
NleDEYSNQNNFVHDCVNITIKQHTVTTTT
KGENFTETDVKNleNleERVVEQNleCITQY
ERESQAYYQRGSS
MoxSKKRPKPGGWNTGGSRYPGQGSP
GGNRYPPQGGGGWGQPHGGGWGQP
HGGGWGQPHGGGWGQPHGGGWGQ
GGGTHSQWNKPSKPKTNMoxKHMoxAG
AAAAGAVVGGLGGYMoxLGSAMoxSRPII
HFGSDYEDRYYRENMoxHRYPNQVYYRP
MoxDEYSNQNNFVHDCVNITIKQHTVTTTT
KGENFTETDVKMoxMoxERVVEQMoxCITQY
ERESQAYYQRGSS
S
H2N
COOH
Met
H2N
COOH
Nle
O
H2N
COOH
Mox
S
H2N
COOH
Met
PrP^C
?

Fig. 36 Chemical model for prion protein conversion due to changes of the conformational preferences of Met side chains. On the top, the amino acid sequence of rhPrP^{C}23-231 and underneath its amino acid sequence is shown. On the left Met is exchange against Nle, where on the right Met is exchanged by Mox. The structures below represent the PrP^{C} structure with incorporated Nle (same secondary structure like the wild type), whereas the structure on the left could be the one for Met → Mox exchange, but the correct structure is not known. The structures are adapted from www.bseinfo.org/sciePrionsandDisease.aspx.

Discussion

8.6 Proof of principle of the newly developed chemical model

The first step to test for the accuracy of our chemical model was the secondary structure observation of our new generated prion protein variants. Far-UV CD spectra of Met-rhPrP^C and Nle-rhPrP^C exhibit the typical pattern for α + β proteins, with two characteristic minima at 222 nm and at 208 nm of similar intensity (Fig. 30). However, a stronger dichroic signal of Nle-rhPrP^C might indicate additional protein stabilization as confirmed by secondary structure changes as a function of temperature increase. Interestingly, at 60 °C the secondary structure of Met-rhPrP^C was partly unfolded, where in contrast the secondary structure of Nle-rhPrP^C seemed to be unchanged (Fig. 31 B). Expectedly, this stabilizing effect could as well be observed in the dramatic shift of the T_m value of about 10 °C (Fig. 32 A), by thermal denaturation of the proteins.

Conversely, replacing the Met residues in rhPrP^C by Mox a drastic change in the overall spectral intensity as well as in curve shape was observed. Obviously, the secondary structure of Mox-rhPrP^C is quite different indicating predominant β-sheet configuration (Fig. 30), with a cold denaturation like melting curve (Fig. 32 B).

The second step to test the functionality of our newly developed chemical model was to perform *de novo* aggregation assays. The simulated Met oxidation effect can be seen in Fig. 33 A, where the aggregation level of Mox-rhPrP^C was three times higher than that of Met-rhPrP^C. Moreover, it was possible with the Nle-rhPrP^C variant to generate a protein that inhibits protein aggregation, since the aggregation level was reduced to about 50% compared to Met-rhPrP^C. In order to assess whether the opposite aggregation tendency of the Mox-rhPrP^C and Nle-rhPrP^C variants is caused directly by the incorporation of the two Met analogs we performed additional *de novo* aggregation assays with increasing $NaIO_4$ concentrations. The finding that the Nle variant showed only a slight increase in the aggregation tendency indicates that the comparably higher aggregation tendency of the parent protein results from the oxidation of Met residues (Fig. 33 B). Additionally, the aggregation level of the Mox variant remained unaltered during $NaIO_4$ addition (Fig. 33 B). This indicates, as aspected, that the oxidant does not have any effect on the Mox variant. Therefore,

Discussion

the high aggregation tendency of Mox-rhPrP^C is caused by Mox which mimics Met(O).

The Mox-rhPrP^C and Nle-rhPrP^C variants represent a reliable model that clearly demonstrates how the hydrophilicity/hydrophobicity of the side chains in structural positions occupied by Met is crucial for the α → β conversion within the rhPrP^C structure. Additionally, we proved that anti- and pro-aggregation prion proteins can be generated by substituting the Met residues with analogs of opposite hydropathy.

8.7 The model peptide

Incorporation of Mox and Nle had dramatic effects on rhPrP^C structure as we observed via CD spectroscopy and *de novo* aggregation assay. In order to be sure that these dramatic effects were caused by the hydrophilicity/hydrophobicity of the two Met analog side chains we assessed their impact on the structure of a suitable model peptide Ac-YLKA**M**LEA**M**AKL**M**AKL**M**A-NH$_2$ described by Dado and Gellman (66). This peptide is known to undergo an α → β transition upon Met-oxidation. From peptide studies Met is well known to stabilize and efficiently induce α helical conformations (108, 109). Similarly, Nle exhibits an even higher intrinsic preference for α-helical states but lower preferences for β-sheet conformations than Met residues (110). No experimental data on structural propensities of Mox residues are available, but it is conceivable that it prefers β-sheet over α-helical conformation because of its hydrophilicity.

The Nle-peptide exhibits the typical α-helical CD profile with a very high content of ordered structure (> 80% α-helix). Conversely, the related Mox-peptide shows a dramatically decreased α-helical content (~ 40%) with a 6-fold increase in random coil, but also a significant percentage of β-type structure (20 %) (Fig. 34). These results are in full agreement with our observations with the Met-rhPrP^C variants.

In fact, model studies on the peptides related to α -helix II of PrP^C clearly revealed a surprisingly low free energy difference between the α-helical and β-sheet conformations of only 5-8 kJmol^{-1} confirming at least for this helix a conformational ambivalence (126, 127).

Discussion

Taken together, one can conclude that Mox represents a good mimic for Met(O). Nonetheless, we have to be aware that though Mox represents a perfect surrogate for Met(O), the effect of a Met → Met(O) exchange on rhPrP^C aggregation propensity as well as secondary structure might have been even more dramatic possibly due to steric effects in addition to the hydropathy change. From our results we strongly believe that Met oxidation in the prion protein can induce the α → β transition, which further leads to aggregation and finally to disease.

9 Conclusions

The conversion process of PrP^C into its pathological isoform PrP^{Sc} is considered to play a central role in prion associated disorders (71). Therefore *in vitro* transitions of recombinant PrP into β-sheet enriched, aggregated structures, which mimic the characteristics of infectious PrP^{Sc} is of fundamental importance to elucidate the initial event in pathogenesis of prion diseases. Although the initial onset for the conformational change is not known, there is increasing evidence that oxidative stress plays an important role in prion infection. Moreover, ROS toxicity has been implicated in several other neurological disorders as well as in aging (83). In this context, the amino acid Met might be of particular importance, since it is readily oxidized by most reactive species. In addition, the oxididation of Met side chains changed the conformational preferences in model peptides (66). In the PrP^C structure, the Met residues are playing a crucial role in forming the hydrophobic core of the globular (i.e. ordered) domain. Therefore, it is conceivable that oxidation of these residues in the hydrophobic core destroys the amphipathic nature of the α-helixes. In other words, the structural α → β transition of PrP^C is directly related to the conformational preferences of Met and Met(O) residues. In fribrillar PrP^{Sc}, such initial chemical modification may be difficult to detect, as a small seed of modified molecules may suffice for the onset of the process.

In experimental Met oxidation studies of PrP^C, controllable and selective side chain oxidation is difficult to achieve. Hence to study the role of Met oxidation in the full length rhPrP^C(23–231) and therefore the α → β structural conversion, the residue-specific incorporation of Nle and Mox provided a useful chemical model. The results we obtained with our experimental model strongly support a natural scenario where oxidative conversion of Met residues, especially buried ones, into Met(O) may act as primary event in sporadic prion disease induction, particularly taking into account the pronounced structural ambivalence of PrP^C.

The non-canonical amino acid incorporation of Nle and Mox provides a tool to generate prion proteins that are arrested in conformational states and mimic PrP^C or even PrP^{Sc}. Thus, this approach might in general be a suitable tool to explore the

Conclusions

underlying mechanisms of crucial biological events, such as structural transitions in many other pathophysiologically relevant proteins. This is demonstrated by an excellent correlation between solution properties of Met/Nle/Mox side chains and the conformational states of the related protein variants. Not surprisingly, this enabled us to anticipate Met oxidation as an initial destabilizing event of the rhPrP^C α-fold and its subsequent transition and assembly into rhPrP^{Sc}.

Finally, recent evidence indicates that diverse neurodegenerative diseases might have a common cause and pathological mechanism: first, the misfolding, second, the aggregation and finally, the accumulation of proteins in the brain resulting in neuronal apoptosis (64). Moreover, several studies coming from different research fields, as well as distinct diseases, strongly support this hypothesis and suggest that a common therapy for these devastating disorders might be possible. Consequently, detailed understanding of the very early steps in the pathogenesis of these diseases will certainly help to elucidate the crucial events in the disease process, towards which suitable therapeutic strategies can be directed. In this context, it is conceivable that the experimental approach, which was developed in this study, will be of prime importance for studies of other proteins, highly relevant for neurodegenerative as well as other aging-related diseases.

10 References

1. Miescher F (1871) Über Die Chemische Zusammensetzung Der Eiterzellen. *Hoppe-Seyler´s medizinisch-chemische Untersuchungen* 4, 441-460.

2. Avery OT, MacLeod CM & McCarty M (1944) Studies on the Chemical Nature of the Substance Inducing Transformation of Pneumococcal Types: Induction of Transformation by a DNA Fraction Isolated from Pneumococcus Type III. *J. Exp. Med.* 79, 137-158.

3. Franklin RE & Gosling RG (1953) Evidence for 2-Chain Helix in Crystalline Structure of Sodium Deoxyribonucleate. *Nature* 172, 156-157.

4. Franklin RE & Gosling RG (1953) Molecular Configuration in Sodium Thymonucleate. *Nature* 171, 740-741.

5. Watson JD & Crick FHC (1953) The Structure of DNA. *Cold Spring Harb Symp Quant Biol* 18, 123-131.

6. Watson JD & Crick FHC (1953) Molecular Structure of Nucleic Acids: A Structure for Deoxyribose Nucleic Acid. *Nature* 171, 737-738.

7. Watson JD & Crick FHC (1953) Genetical Implications of the Structure of Deoxyribonucleic Acid. *Nature* 171, 964-967.

8. Crick FH (1957) On Protein Synthesis. *Symp Soc Exp Biol* 12, 138-63.

9. Stryer L, Tymoczko JL & Berg JM (2007) Stryer Biochemie (Elsevier GmbH, Spektrum Akademischer Verlag).

10. Matthaei H & Nirenberg MW (1961) Dependence of Cell-Free Protein Synthesis in E Coli Upon RNA Prepared from Ribosomes. *Biochem Biophys Res Commun* 4, 404-&.

11. Knippers R (2006) Molekulare Genetik (Thieme, Stuttgart).

12. Hames D & Hooper N (2005) Biochemistry (Instant Notes) (Taylor & Francis Group, Abingdon).

References

13. Kennell D & Riezman H (1977) Transcription and Translation Initiation Frequencies of the Escherichia Coli Lac Operon. *J Mol Biol* 114, 1-21.

14. Budisa N (2006) Engineering the Genetic Code-Expanding the Amino Acid Repertoire for the Design of Novel Proteins (WILEY-VCH Verlag, Weilheim).

15. Köhrer C & RajBhandary U (2009) Protein Engineering (Springer-Verlag, Berlin).

16. Srinivasan G, James CM & Krzycki JA (2002) Pyrrolysine Encoded by UAG in Archaea: Charging of a UAG-Decoding Specialized tRNA. *Science* 296, 1459-1462.

17. Cowie DB & Cohen GN (1957) Biosynthesis by Escherichia Coli of Active Altered Proteins Containing Selenium Instead of Sulfur. *Biochim Biophys Acta* 26, 252-261.

18. Hendrickson WA, Horton JR & Lemaster DM (1990) Selenomethionyl Proteins Produced for Analysis by Multiwavelength Anomalous Diffraction (Mad) - a Vehicle for Direct Determination of 3-Dimensional Structure. *EMBO J* 9, 1665-1672.

19. Cohen GN & Munier R (1959) Effets Des Analogues Structuraux D'aminoacids Sur La Croissance, La Synthèse De Protéines Et La Synthèse D'enzymes Chez Escherichia Coli. *Biochim Biophys Acta* 31, 347-356.

20. Budisa N, Minks C, Alefelder S, Wenger W, Dong F, *et al.* (1999) Toward the Experimental Codon Reassignment in Vivo: Protein Building with an Expanded Amino Acid Repertoire. *FASEB J* 13, 41-51.

21. Weber P, Reznicek L, Mitteregger G, Kretzschmar H & Giese A (2008) Differential Effects of Prion Particle Size on Infectivity in Vivo and in Vitro. *Biochem Biophys Res Commun* 369, 924-928.

22. O'Brien PJ & Herschlag D (1999) Catalytic Promiscuity and the Evolution of New Enzymatic Activities. *Chem Biol* 6, R91-R105.

References

23. Hartman MCT, Josephson K & Szostak JW (2006) Enzymatic Aminoacylation of tRNA with Unnatural Amino Acids. *Proc Natl Acad Sci USA* 103, 4356-4361.

24. Hartman MCT, Josephson K, Lin C-W & Szostak JW (2007) An Expanded Set of Amino Acid Analogs for the Ribosomal Translation of Unnatural Peptides. *PLoS ONE* 2, e972.

25. Fersht AR & Dingwall C (1979) Editing Mechanism for the Methionyl-Transfer RNA-Synthetase in the Selection of Amino-Acids in Protein-Synthesis. *Biochemistry* 18, 1250-1256.

26. Rose GD & Wolfenden R (1993) Hydrogen-Bonding, Hydrophobicity, Packing, and Protein-Folding. *Annu Rev Biophys Biomol Struct* 22, 381-415.

27. Rubini M, Lepthien S, Golbik R & Budisa N (2006) Aminotryptophan-Containing Barstar: Structure-Function Tradeoff in Protein Design and Engineering with an Expanded Genetic Code. *Biochim Biophys Acta* 1764, 1147-1158.

28. Chapeville F, Lipmann F, von Ehrenstein G, Weisblum B, Ray WJ, *et al.* (1962) On the Role of Soluble RNA in Coding for Amino Acids. *Proc Natl Acad Sci USA* 48, 1086-1092.

29. Noren CJ, Anthony-Cahill SJ, Griffith MC & Schultz PG (1989) A General Method for Site-Specific Incorporation of Unnatural Amino Acids into Proteins. *Science* 244, 182-188.

30. Arslan T, Mamaev SV, Mamaeva NV & Hecht SM (1997) Structurally Modified Firefly Luciferase. Effects of Amino Acid Substitution at Position 286. *J Am Chem Soc* 119, 10877-10887.

31. Bain JD, Diala ES, Glabe CG, Dix TA & Chamberlin AR (1989) Biosynthetic Site-Specific Incorporation of a Non-Natural Amino Acid into a Polypeptide. *J Am Chem Soc* 111, 8013-8014.

32. Mendel D, Cornish VW & Schultz PG (1995) Site-Directed Mutagenesis with an Expanded Genetic Code. *Annu Rev Biophys Biomol Struct* 24, 435-462.

References

33. Kwon I, Kirshenbaum K & Tirrell DA (2003) Breaking the Degeneracy of the Genetic Code. *J Am Chem Soc* 125, 7512-7513.

34. Hohsaka T, Kajihara D, Ashizuka Y, Murakami H & Sisido M (1999) Efficient Incorporation of Nonnatural Amino Acids with Large Aromatic Groups into Streptavidin in in Vitro Protein Synthesizing Systems. *J Am Chem Soc* 121, 34-40.

35. Breydo L, Bocharova OV, Makarava N, Salnikov VV, Anderson M, et al. (2005) Methionine Oxidation Interferes with Conversion of the Prion Protein into the Fibrillar Proteinase K-Resistant Conformation. *Biochemistry* 44, 15534-15543.

36. Tamura K & Schimmel P (2002) Ribozyme Programming Extends the Protein Code. *Nat Biotech* 20, 669-670.

37. Bain JD, Switzer C, Chamberlin R & Bennert SA (1992) Ribosome-Mediated Incorporation of a Non-Standard Amino Acid into a Peptide through Expansion of the Genetic Code. *Nature* 356, 537-539.

38. Rackham O & Chin JW (2005) A Network of Orthogonal Ribosome-mRNA Pairs. *Nat Chem Biol* 1, 159-166.

39. Wang K, Neumann H, Peak-Chew SY & Chin JW (2007) Evolved Orthogonal Ribosomes Enhance the Efficiency of Synthetic Genetic Code Expansion. *Nat Biotech* 25, 770-777.

40. Sisido, Sisido M, Hohsaka & Hohsaka T (2001) Introduction of Specialty Functions by the Position-Specific Incorporation of Nonnatural Amino Acids into Proteins through Four-Base Codon/Anticodon Pairs. *Appl Microbiol Biotechnol* 57, 274-281.

41. Hohsaka T, Ashizuka Y, Taira H, Murakami H & Sisido M (2001) Incorporation of Nonnatural Amino Acids into Proteins by Using Various Four-Base Codons in an Escherichia Coli in Vitro Translation System. *Biochemistry* 40, 11060-11064.

References

42. Magliery TJ, Anderson JC & Schultz PG (2001) Expanding the Genetic Code: Selection of Efficient Suppressors of Four-Base Codons and Identification of "Shifty" Four-Base Codons with a Library Approach in Escherichia Coli. *J Mol Biol* 307, 755-769.

43. Hohsaka T & Sisido M (2002) Incorporation of Non-Natural Amino Acids into Proteins. *Curr Opin Chem Biol* 6, 809-815.

44. Budisa N (2004) Prolegomena to Future Experimental Efforts on Genetic Code Engineering by Expanding Its Amino Acid Repertoire. *Angew Chem Int Ed Engl* 43, 6426-6463.

45. Vogt W (1995) Oxidation of Methionyl Residues in Proteins - Tools, Targets, and Reversal. *Free Radic. Biol. Med.* 18, 93-105.

46. Brosnan JT & Brosnan ME (2006) The Sulfur-Containing Amino Acids: An Overview. *J. Nutr.* 136, 1636S-1640.

47. Voet D, Voet JG & Pratt CW (2008) Fundamentals of Biochemistry (John Wiley & Sons, Inc., Hoboken).

48. Drabkin HJ, Estrella M & Rajbhandary UL (1998) Initiator-Elongator Discrimination in Vertebrate tRNAs for Protein Synthesis. *Mol Cell Biol* 18, 1459-1466.

49. Cantoni GL (1953) S-Adenosylmethionine; a New Intermediate Formed Enzymatically from L-Methionine and Adenosinetriphosphate. *J Biol Chem* 204, 403-416.

50. Brosnan JT, Brosnan ME, Bertolo RFP & Brunton JA (2007) Methionine: A Metabolically Unique Amino Acid. *Livest Sci* 112, 2-7.

51. Lagerwerf FM, van de Weert M, Heerma W & Haverkamp J (1996) Identification of Oxidized Methionine in Peptides. *Rapid Commun. Mass Spectrom.* 10, 1905-1910.

52. Levine RL, Moskovitz J & Stadtman ER (2000) Oxidation of Methionine in Proteins: Roles in Antioxidant Defense and Cellular Regulation. *Iubmb Life* 50, 301-307.

References

53. Davies MJ (2005) The Oxidative Environment and Protein Damage. *Biochim Biophys Acta* 1703, 93-109.

54. Petropoulos I & Friguet B (2005) Protein Maintenance in Aging and Replicative Senescence: A Role for the Peptide Methionine Sulfoxide Reductases. *Biochim Biophys Acta* 1703, 261-266.

55. Moskovitz J (2005) Methionine Sulfoxide Reductases: Ubiquitous Enzymes Involved in Antioxidant Defense, Protein Regulation, and Prevention of Aging-Associated Diseases. *Biochim Biophys Acta* 1703, 213–219.

56. Moskovitz J (2005) Roles of Methionine Suldfoxide Reductases in Antioxidant Defense, Protein Regulation and Survival. *Curr Pharm Des* 11, 1451-1457.

57. Moskovitz J, Bar-Noy S, Williams WM, Berlett BS & Stadtman ER (2001) Methionine Sulfoxide Reductase (MsrA) Is a Regulator of Antioxidant Defense and Lifespan in Mammals. *Proc Natl Acad Sci USA* 98, 12920-12925.

58. Ruan H, Tang XD, Chen ML, Joiner MA, Sun G, et al. (2002) High-Quality Life Extension by the Enzyme Peptide Methionine Sulfoxide Reductase. *Proc Natl Acad Sci USA* 99, 2748-2753.

59. Dobson CM (2005) An Overview of Protein Misfolding Diseases (Wiley-VCH Verlag GmbH & Co. KGaA, Weinheim).

60. Anfinsen CB, Redfield RR, Choate WL, Page J & Carroll WR (1954) Studies on the Gross Structure, Cross-Linkages, and Terminal Sequences in Ribonuclease. *J Biol Chem* 207, 201-210.

61. Dobson CM (2006) Protein Aggregation and Its Consequences for Human Disease. *Protein Pept. Lett.* 13, 219-227.

62. Chiti F & Dobson CM (2006) Protein Misfolding, Functional Amyloid, and Human Disease. *Annu Rev Biochem* 75, 333-366.

63. Soto C (2003) Unfolding the Role of Protein Misfolding in Neurodegenerative Diseases. *Nat Rev Neurosci* 4, 49-60.

References

64. Soto C (2006) Prions-the New Biology of Proteins (Taylor & Francis Group, Boca Raton).

65. Ehud G (2002) The 'Correctly Folded' State of Proteins: Is It a Metastable State? *Angew Chem Int Ed Engl* 41, 257-259.

66. Dado GP & Gellman SH (1993) Redox Control of Secondary Structure in a Designed Peptide. *J Am Chem Soc* 115, 12609-12610.

67. Dayhoff MO, Barker WC & Hunt LT (1983) Establishing Homologies in Protein Sequences. *Methods Enzymol* 91, 524-545.

68. Petropoulos I & Friguet B (2006) Maintenance of Proteins and Aging: The Role of Oxidized Protein Repair. *Free Radic Res* 40, 1269-1276.

69. Belay ED (1999) Transmissible Spongiform Encephalopathies in Humans. *Annu Rev Microbiol* 53, 283-314.

70. Belay ED & Schonberger LB (2005) The Public Health Impact of Prion Diseases 1. *Annu Rev Public Health* 26, 191–212.

71. Stanley B. Prusiner MRS, Stephen J. DeArmond, and Fred E. Cohen (1998) Prion Protein Biology. *Cell* 93, 337–348.

72. Lysek DA, Schorn C, Nivon LG, Esteve-Moya V, Christen B, et al. (2005) Prion Protein NMR Structures of Cats, Dogs, Pigs, and Sheep. *Proc Natl Acad Sci USA* 102, 640-645.

73. Hornemann S, Korth C, Oesch B, Riek R, Wider G, et al. (1997) Recombinant Full-Length Murine Prion Protein, mPrP(23-231): Purification and Spectroscopic Characterization. *FEBS Lett* 413, 277-281.

74. Donne DG, Viles JH, Groth D, Mehlhorn I, James TL, et al. (1997) Structure of the Recombinant Full-Length Hamster Prion Protein PrP(29-231): The N Terminus Is Highly Flexible. *Proc Natl Acad Sci USA* 94, 13452-13457.

75. Weissmann C (2004) The State of the Prion. *Nat Rev Microbiol* 2, 861-871.

76. Pan K-M, Baldwin M, Nguyen J, Gasset M, Serban A, et al. (1993) Conversion of Alpha-Helices into Beta-Sheets Features in the Formation of the Scrapie Prion Proteins. *Mol Biol Cell* 4, 313A.

References

77. Prusiner SB (1991) Molecular-Biology of Prion Diseases. *Science* 252, 1515-1522.

78. Prusiner SB (1997) Prion Diseases and the BSE Crisis. *Science* 278, 245-251.

79. Mrak RE, Griffin WST & Graham DI (1997) Aging-Associated Changes in Human Brain. *J Neuropathol Exp Neurol* 56, 1269-1275.

80. Martin JB (1999) Molecular Basis of the Neurodegenerative Disorders. *N Engl J Med* 340, 1970-1980.

81. Barnham KJ, Cappai R, Beyreuther K, Masters CL & Hill AF (2006) Delineating Common Molecular Mechanisms in Alzheimer's and Prion Diseases. *Trends Biochem Sci* 31, 465-472.

82. Bieschke J, Weber P, Sarafoff N, Beekes M, Giese A, *et al.* (2004) Autocatalytic Self-Propagation of Misfolded Prion Protein. *Proc Natl Acad Sci USA* 101, 12207-12211.

83. Milhavet O, McMahon HEM, Rachidi W, Nishida N, Katamine S, *et al.* (2000) Prion Infection Impairs the Cellular Response to Oxidative Stress. *Proc Natl Acad Sci USA* 97, 13937-13942.

84. DeMarco ML & Daggett V (2005) Local Environmental Effects on the Structure of the Prion Protein. *Comptes Rendus Biologies* 328, 847-862.

85. Prusiner SB (1998) Prions. *Proc Natl Acad Sci USA* 95, 13363-13383.

86. Budisa N, Steipe B, Demange P, Eckerskorn C, Kellermann J, *et al.* (1995) High-Level Biosynthetic Substitution of Methionine in Proteins by Its Analogs 2-Aminohexanoic Acid, Selenomethionine, Telluromethionine and Ethionine in Escherichia-Coli. *Eur J Biochem* 230, 788-796.

87. Giese A, Levin J, Bertsch U & Kretzschmar H (2004) Effect of Metal Ions on De Novo Aggregation of Full-Length Prion Protein. *Biochem Biophys Res Commun* 320, 1240-1246.

References

88. Hornemann S & Glockshuber R (1996) Autonomous and Reversible Folding of a Soluble Amino-Terminally Truncated Segment of the Mouse Prion Protein. *J Mol Biol* 261, 614-619.

89. Merkel L, Cheburkin Y, Wiltschi B & Budisa B (2007) In Vivo Chemoenzymatic Control of N-Terminal Processing in Recombinant Human Epidermal Growth Factor. *Chembiochem* 8, 2227-2232.

90. Varshavsky A (1992) The N-End Rule. *Cell* 69, 725-735.

91. Schwille P, MeyerAlmes FJ & Rigler R (1997) Dual-Color Fluorescence Cross-Correlation Spectroscopy for Multicomponent Diffusional Analysis in Solution. *Biophys J* 72, 1878-1886.

92. Liemann S & Glockshuber R (1999) Influence of Amino Acid Substitutions Related to Inherited Human Prion Diseases on the Thermodynamic Stability of the Cellular Prion Protein. *Biochemistry* 38, 3258-3267.

93. Gill SC & von Hippel PH (1989) Calculation of Protein Extinction Coefficients from Amino Acid Sequence Data. *Anal Biochem* 182, 319-326.

94. Foster LJ, de Hoog CL & Mann M (2003) Unbiased Quantitative Proteomics of Lipid Rafts Reveals High Specificity for Signaling Factors. *Proc Natl Acad Sci USA* 100, 5813-5818.

95. Olsen JV, de Godoy LMF, Li G, Macek B, Mortensen P, *et al.* (2005) Parts Per Million Mass Accuracy on an Orbitrap Mass Spectrometer Via Lock Mass Injection into a C-Trap. *Mol Cell Proteomics* 4, 2010-2021.

96. Gruhler A, Olsen JV, Mohammed S, Mortensen P, Faergeman NJ, *et al.* (2005) Quantitative Phosphoproteomics Applied to the Yeast Pheromone Signaling Pathway. *Mol Cell Proteomics* 4, 310-327.

97. Knowles JR (1965) Role of Methionine in Alpha-Chymotrypsin-Catalysed Reactions. *Biochem J* 95, 180-&.

References

98. Bieschke J, Giese A, Schulz-Schaeffer W, Zerr I, Poser S, et al. (2000) Ultrasensitive Detection of Pathological Prion Protein Aggregates by Dual-Color Scanning for Intensely Fluorescent Targets. *Proc Natl Acad Sci USA* 97, 5468-5473.

99. Giese A, Bieschke J, Eigen M & Kretzschmar HA (2000) Putting Prions into Focus: Application of Single Molecule Detection to the Diagnosis of Prion Diseases. *Arch Virol*, 161-171.

100. Schwille P, Bieschke J & Oehlenschlager F (1997) Kinetic Investigations by Fluorescence Correlation Spectroscopy: The Analytical and Diagnostic Potential of Diffusion Studies. *Biophys Chem* 66, 211-228.

101. Post K, Pitschke M, Schafer O, Wille H, Appel TR, et al. (1998) Rapid Acquisition of Beta-Sheet Structure in the Prion Protein Prior to Multimer Formation. *Biol Chem* 379, 1307-1317.

102. Requenaa JR, Dimitrovaa MN, Legnameb G, Teijeirae S, Prusiner SB, et al. (2004) Oxidation of Methionine Residues in the Prion Protein by Hydrogen Peroxide. *Arch Biochem Biophys* 432 188–195.

103. Redecke L, von Bergen M, Clos J, Konarev PV, Svergun DI, et al. (2007) Structural Characterization of Beta-Sheeted Oligomers Formed on the Pathway of Oxidative Prion Protein Aggregation in Vitro. *J Struct Biol* 157, 308-320.

104. Canello T, Engelstein R, Moshel O, Xanthopoulos K, Juanes ME, et al. (2008) Methionine Sulfoxides on PrP^{Sc}: A Prion-Specific Covalent Signature. *Biochemistry* 47, 8866-8873.

105. Budisa N, Huber R, Golbik R, Minks C, Weyher E, et al. (1998) Atomic Mutations in Annexin V - Thermodynamic Studies of Isomorphous Protein Variants. *Eur J Biochem* 253, 1-9.

106. Privalov PL (1990) Cold Denaturation of Proteins. *Biophys J* 57, A26-A26.

107. Blandamer MJ, Briggs B, Burgess J & Cullis PM (1990) Thermodynamic Model for Isobaric Heat-Capacities Associated with Cold and Warm Thermal-Denaturation of Proteins. *J Chem Soc-Faraday Trans* 86, 1437-1441.

References

108. Creamer TP & Rose GD (1994) Alpha-Helix-Forming Propensities in Peptides and Proteins. *Proteins* 19, 85-97.

109. Minor DL & Kim PS (1994) Measurement of the Beta-Sheet-Forming Propensities of Amino-Acids. *Nature* 367, 660-663.

110. Lyu PC, Sherman JC, Chen A & Kallenbach NR (1991) Alpha-Helix Stabilization by Natural and Unnatural Amino-Acids with Alkyl Side-Chains. *Proc Natl Acad Sci USA* 88, 5317-5320.

111. Hornemann S, Schorn C & Wuthrich K (2004) NMR Structure of the Bovine Prion Protein Isolated from Healthy Calf Brains. *EMBO Rep.* 5, 1159-1164.

112. Leclerc E, Peretz D, Ball H, Solforosi L, Legname G, et al. (2003) Conformation of PrP^C on the Cell Surface as Probed by Antibodies. *J Mol Biol* 326, 475-483.

113. Mari L. DeMarco VD (2009) Characterization of Cell-Surface Prion Protein Relative to Its Recombinant Analogue: Insights from Molecular Dynamics Simulations of Diglycosylated, Membrane-Bound Human Prion Protein. *J Neurochem* 9999.

114. Parchi P, Capellari S, Chen SG, Petersen RB, Gambetti P, et al. (1997) Typing Prion Isoforms. *Nature* 386, 232-233.

115. Elfrink K, Ollesch J, Stöhr J, Willbold D, Riesner D, et al. (2008) Structural Changes of Membrane-Anchored Native Prp^c. *Proc Natl Acad Sci USA* 105, 10815-10819.

116. Levine M & Tarver H (1951) Studies on Ethionine III. Incorporation of Ethionine into Rat Proteins. *J Biol Chem* 192, 835-850.

117. Linton S, Davies MJ & Dean RT (2001) Protein Oxidation and Ageing. *Exp Gerontol* 36, 1503-1518.

118. Langedijk JPM, Fuentes G, Boshuizen R & Bonvin AMJJ (2006) Two-Rung Model of a Left-Handed [Beta]-Helix for Prions Explains Species Barrier and Strain Variation in Transmissible Spongiform Encephalopathies. *J Mol Biol* 360, 907-920.

References

119. Palmer MS, Dryden AJ, Hughes JT & Collinge J (1991) Homozygous Prion Protein Genotype Predisposes to Sporadic Creutzfeldt-Jakob Disease. *Nature* 352, 340-342.

120. Lewis PA, Tattum MH, Jones S, Bhelt D, Batchelor M, *et al.* (2006) Codon 129 Polymorphism of the Human Prion Protein Influences the Kinetics of Amyloid Formation. *J Gen Virol* 87, 2443-2449.

121. Winklhofer KF, Heske J, Heller U, Reintjes A, Muranyi W, *et al.* (2003) Determinants of the in Vivo Folding of the Prion Protein - a Bipartite Function of Helix 1 in Folding and Aggregation. *J Biol Chem* 278, 14961-14970.

122. Hirschberger T, Stork M, Schropp B, Winklhofer KF, Tatzelt J & Tavan P (2006) Structural Instability of the Prion Protein Upon M205S/R Mutations Revealed by Molecular Dynamics Simulations. *Biophys J* 90, 3908–3918

123. Colombo G, Meli M, Morra G, Gabizon R & Gasset Ma (2009) Methionine Sulfoxides on Prion Protein Helix-3 Switch on the α-Fold Destabilization Required for Conversion. *PLoS ONE* 4, e4296.

124. Ross CA & Poirier MA (2005) What Is the Role of Protein Aggregation in Neurodegeneration? *Nat Rev Mol Cell Biol* 6, 891-898.

125. Jansen K, Schafer O, Birkmann E, Post K, Serban H, *et al.* (2001) Structural Intermediates in the Putative Pathway from the Cellular Prion Protein to the Pathogenic Form. *Biol Chem* 382, 683-691.

126. Ronga L, Tizzano B, Palladino P, Ragone R, Urso E, *et al.* (2006) The Prion Protein: Structural Features and Related Toxic Peptides. *Chem Biol Drug Des* 68, 139-147.

127. Ronga L, Palladino P, Saviano G, Tancredi T, Benedetti E, *et al.* (2008) Structural Characterization of a Neurotoxic Threonine-Rich Peptide Corresponding to the Human Prion Protein Alpha2-Helical 180-195 Segment, and Comparison with Full-Length Alpha2-Helix-Derived Peptides. *J Pept Sci* 14, 1096-1102.

Figures and Tables

11 Figure List

Fig. 1 The standard genetic code (RNA format). 6

Fig. 2 Basic prerequisites for noncanonical amino acid incorporation 12

Fig. 3 The four sulfur containing amino acids 15

Fig. 4 Met metabolism 17

Fig. 5 Met oxidation to Met(O) and Met(O_2) 18

Fig. 6 Oxidation-reduction cycle 19

Fig. 7 Three-dimensional structure of human PrP^C(125-231) 24

Fig. 8 Structure of PrP^C and speculative structure of PrP^{Sc} 25

Fig. 9 Measuring systems of a dual color FCS reader with a scanning unit for SIFT measurements and sample plate 41

Fig. 10 Flow chart of the labeling procedure 43

Fig. 11 Flow chart of dilution steps for aggregation measurements 46

Fig. 12 Amino acid sequence of PrP23-231 M129. 48

Fig. 13 Coomassie stained SDS-PAGE gels of different purification steps .. 49

Fig. 14 Deconvoluted mass spectrum of Met-rhPrP^C. 50

Fig. 15 Periodate dependent aggregation of Met-rhPrP^C determined by cross-correlation amplitude G(0) 51

Fig. 16 Schematic representation of basic instrumental setup of Orbitrap ESI-MS/MS and ESI-MS 54

Fig. 17 Oxidation level of the Met-rhPrP^C peptides using 5 equiv. $NaIO_4$ 56

Fig. 18 Secondary structure and thermally induced unfolding profile of Met-rhPrP^C 57

Fig. 19 Secondary structure and thermally induced unfolding profile of $NaIO_4$ with 0.5 equiv. treated Met-rhPrP^C 57

Figures and Tables

Fig. 20 Secondary structure and thermally induced unfolding profile of Met-rhPrP^C treated with 5 equiv NaIO$_4$ 58

Fig. 21 Oxidation level of the Met-rhPrP^C peptides using 25 equiv. (soluble fraction) NaIO$_4$ 59

Fig. 22 Secondary structure and thermally induced unfolding profile of Met-rhPrP^C treated with 25 equiv. NaIO$_4$ 60

Fig. 23 Oxidation level of the Met-rhPrP^C peptides using 25 equiv. (pellet fraction) NaIO$_4$ 61

Fig. 24 Graphical representation of the overall oxidation results in Met containing peptides 62

Fig. 25 Deconvoluted mass spectrum of protein sample containing apportion of fully labeled Met(O)-rhPrP^C 63

Fig. 26 Methionine (Met) and its isosteric analogs norleucine (Nle) and methoxinine (Mox). 64

Fig. 29 Coomassie stained SDS-PAGE gels of Met-rhPrP^C and analogs 67

Fig. 30 CD spectra of Met-rhPrP^C and its Nle- and Mox-variants at 37 °C and 0.2 mg/mL in 10 mM MES pH 6.0. 68

Fig. 31 Secondary structure changes as a function of temperature increase, in Met-rhPrP^C and related variants 70

Fig. 32 Thermal denaturation monitored by the changes of dichroic intensities at 222 nm in function of temperature 71

Fig. 33 *In vitro* aggregation of Met-rhPrP^C compared to Nle-rhPrP^C and Mox-rhPrP^C 73

Fig. 34 Sequence and secondary structure of two variants, Nle (blue) and Mox (red), of the 18 amino acid model peptide 74

Fig. 35 Comparison of the Met oxidation level in the Met205/206 peptide, using different NaIO$_4$ equivalents 81

Fig. 36 Chemical model for prion protein conversion due to changes of the conformational preferences of Met side chains 83

Fig. 37 Mass spectra containing peptide M109 105

Fig. 38 Mass spectra containing peptide M112/M129/M134. 106

Fig. 39 Mass spectra containing peptide M112/M129/M134 107

Fig. 40 Mass spectra containing peptide M154 108

Fig. 41 Mass spectrum containing peptide M166 109

Fig. 42 Mass spectrum containing peptide M205/206. 110

Fig. 43 Mass spectra containing peptide M213 111

Fig. 44 Mass spectra containing peptide M109 112

Fig. 45 Mass spectra containing peptide M112/M129/M134 113

Fig. 46 Mass spectra containing peptide M154 and mass spectra containing peptide M166 114

Fig. 47 Mass spectra containing peptide M205/M206 and mass spectra containing peptide M213 115

Fig. 48 Mass spectra containing peptide M109 116

Fig. 49 Mass spectra containing peptide M112/M129/M134 117

Fig. 50 Mass spectra containing peptide M154 and mass spectra containing peptide M166 118

Fig. 51 Mass spectra containing peptide M205/206 and mass spectra containing peptide M213 119

Figures and Tables

12 Table List

Table 1   N values before PK digest. 44

Table 2   N values after PK digest. 45

Table 3   Primary sequences of the generated Met-PrP^C fragments by trypsin digestion. 52

Table 4   Comparison of the two instrumental approaches (Orbitrap ESI-MS/MS and ESI-MS) using Met oxidation in fragment 6 as example. 53

Table 5   Calculated composition of the different secondary structures of the two peptide analogues, Nle (blue) and Mox (red), of the Dado-Gellmann model peptide. 75

13 Appendix

13.1 Mapping NaIO$_4$ induced Met oxidation by mass spectrometry - 5 equivalents NaIO$_4$

Sequence: TNMK→ M109 [unoxidized]
Theortical mass: 493.237

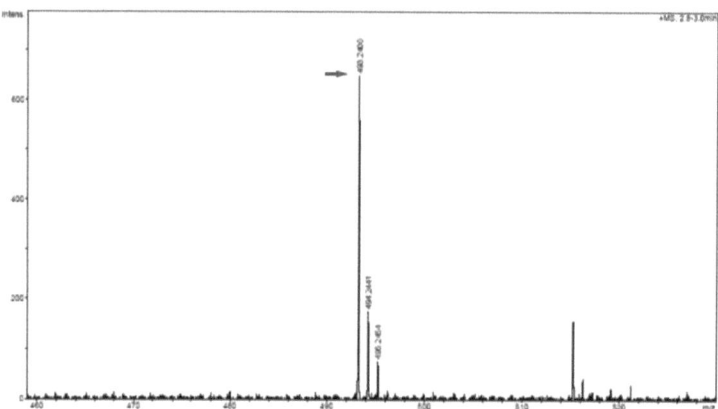

Sequence: TNMK→ M109 [1 x oxidized]
Theortical mass: 509.232

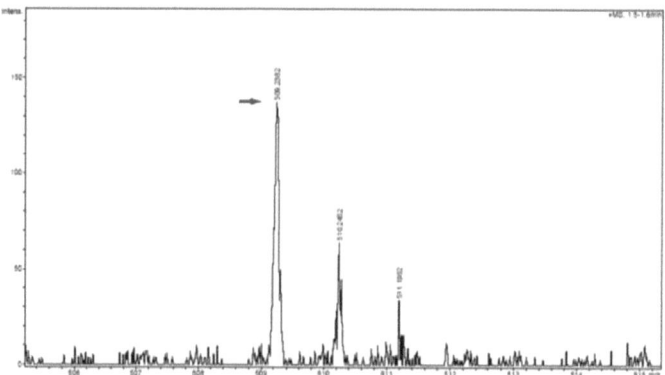

Fig. 37 Mass spectra containing peptide M109. Unoxidized and 1 x oxidized

Appendix

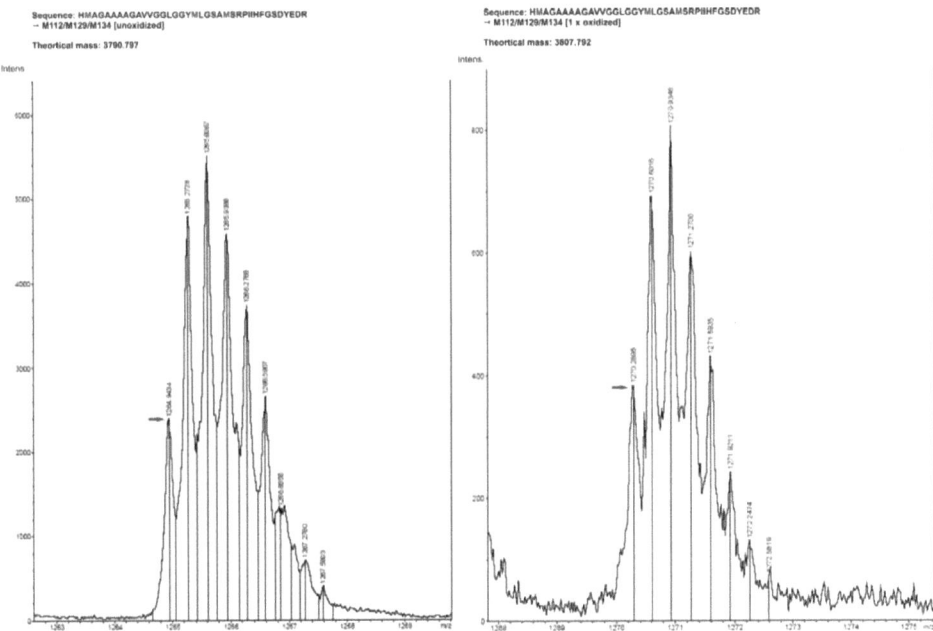

Fig. 38 Mass spectra containing peptide M112/M129/M134. Unoxidized and 1 x oxidized.

Appendix

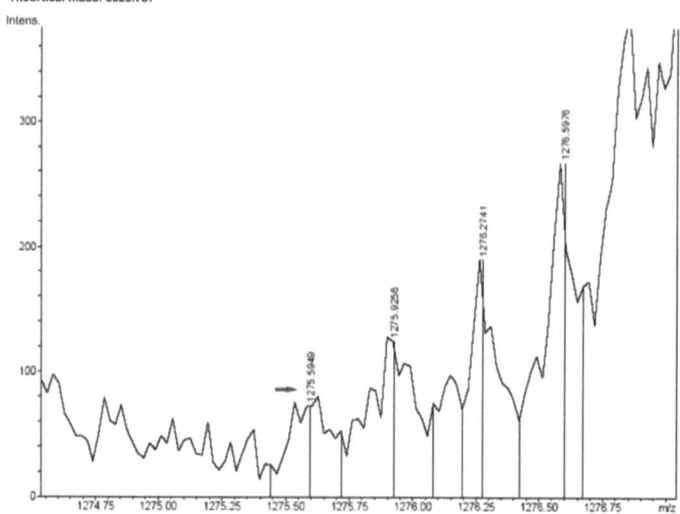

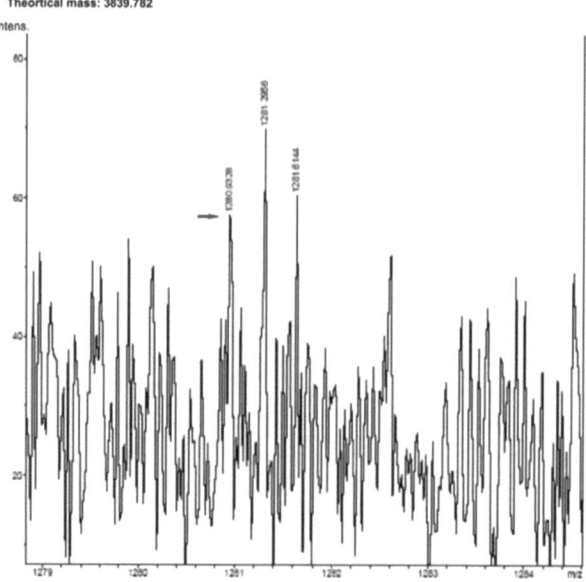

Fig. 39 Mass spectra containing peptide M112/M129/M134. 2 x and 3 x oxidized.

Appendix

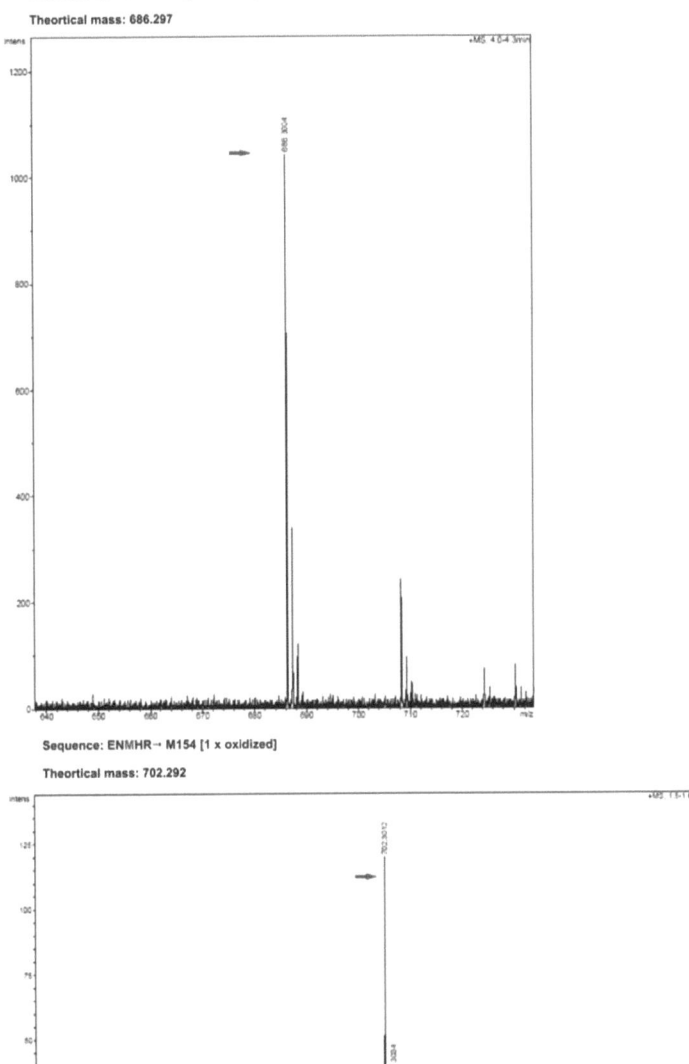

Fig. 40 Mass spectra containing peptide M154. Unoxidized and 1 x oxidized.

Appendix

Sequence: YPNQVYYRPMDEYSNQNNFVHDCVNITIK → M166 [1 x oxidized]

Theortical mass: 3636.640

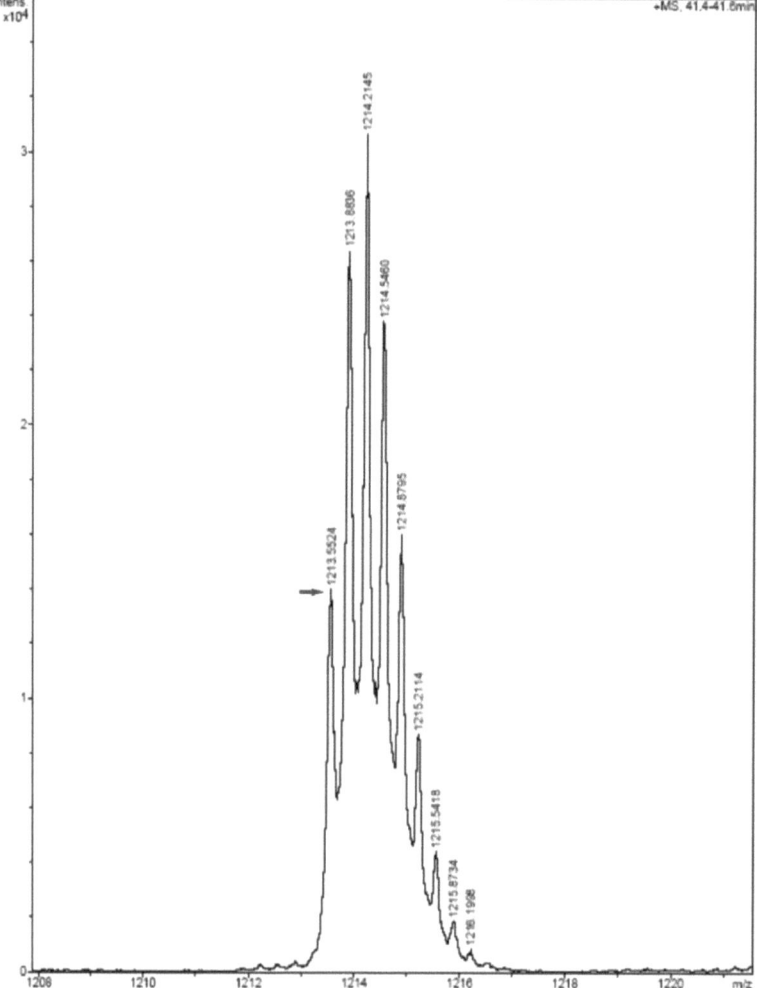

Fig. 41 Mass spectrum containing peptide M166. 1 x oxidized.

Appendix

Sequence: MMER→ M205/M206 [unoxidized]

Theortical mass: 566.235

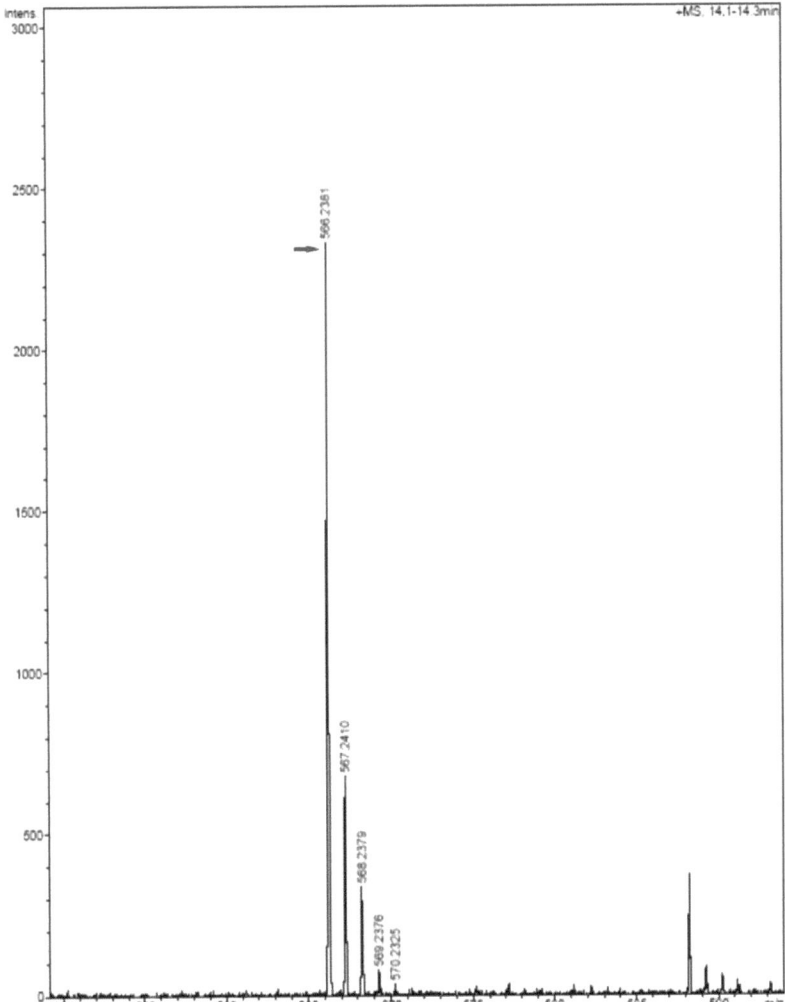

Fig. 42 **Mass spectrum containing peptide M205/206.** Unoxidized.

Appendix

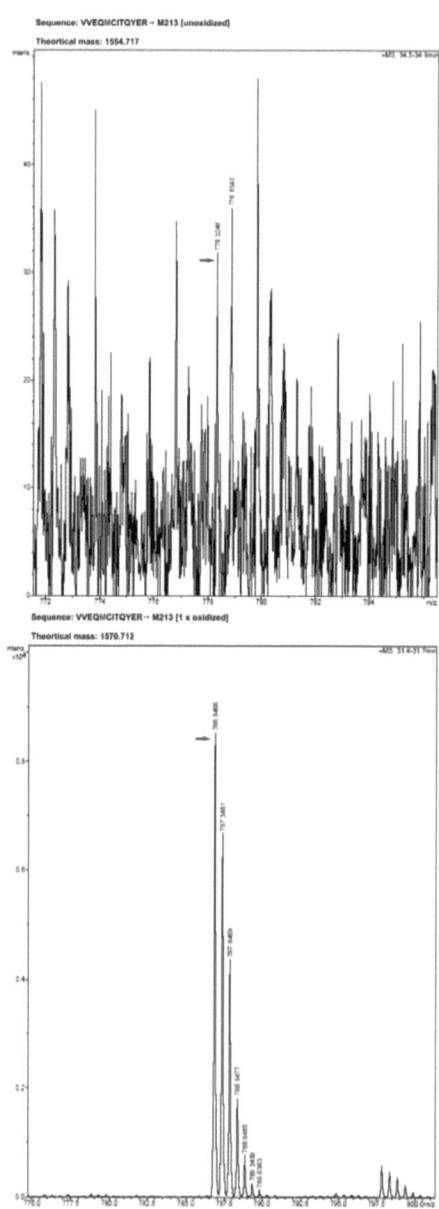

Fig. 43 Mass spectra containing peptide M213. Unoxidized and 1 x oxidized.

13.2 Mapping NaIO₄ induced Met oxidation by mass spectrometry - 25 equivalents NaIO₄- soluble fraction

Sequence: TNMK→ M109 [unoxidized]

Theortical mass: 493.237

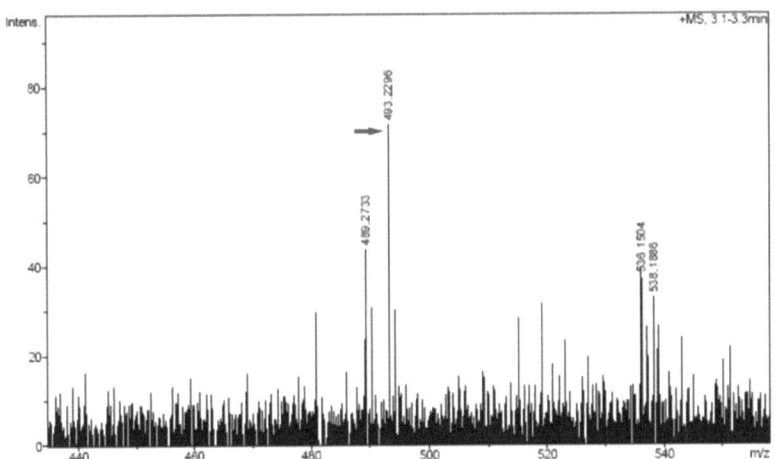

Sequence: TNMK→ M109 [1 x oxidized]

Theortical mass: 509.232

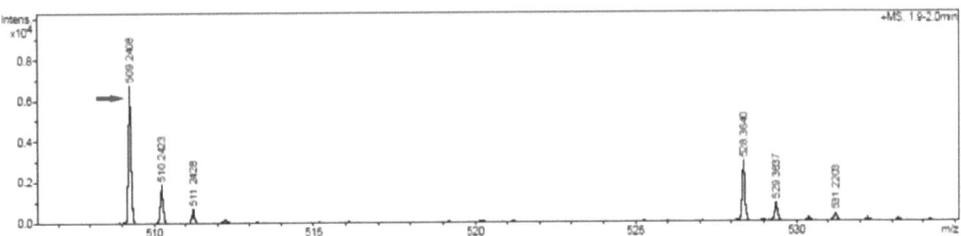

Fig. 44 Mass spectra containing peptide M109. Unoxidized and 1 x oxidized.

Appendix

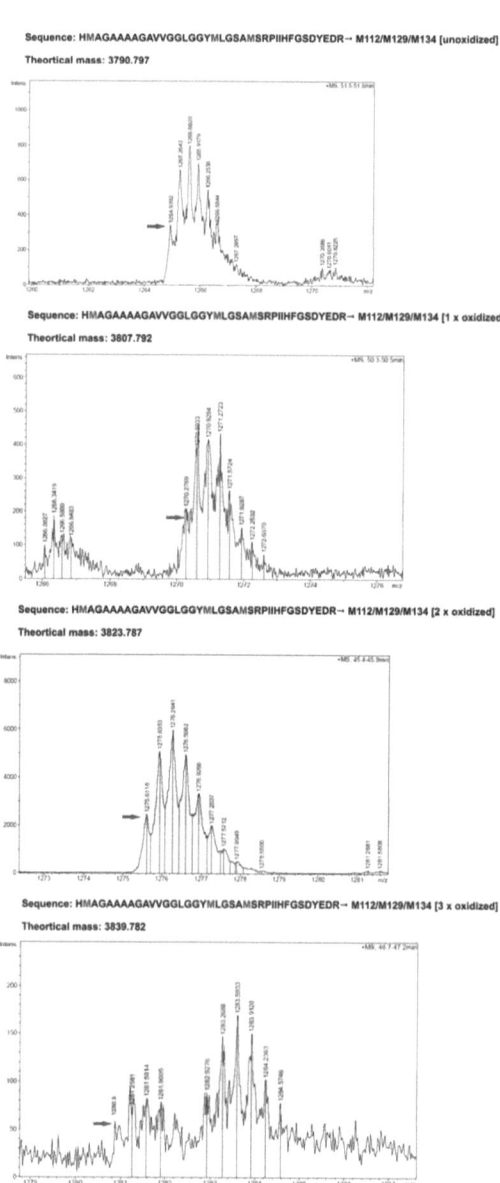

Fig. 45 Mass spectra containing peptide M112/M129/M134. Unoxidized, 1 x, 2 x and 3 x oxidized.

Appendix

Sequence: ENMHR→ M154 [unoxidized]
Theortical mass: 686.297

Sequence: ENMHR→ M154 [1 x oxidized]
Theortical mass: 702.292

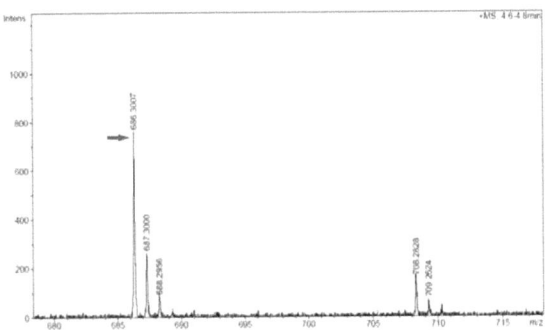

Sequence: YPNQVYYRPMDEYSNQNNFVHDCVNITIK→ M166 [unoxidized]
Theortical mass: 3620.645

Sequence: YPNQVYYRPMDEYSNQNNFVHDCVNITIK→ M166 [1 x oxidized]
Theortical mass: 3636.640

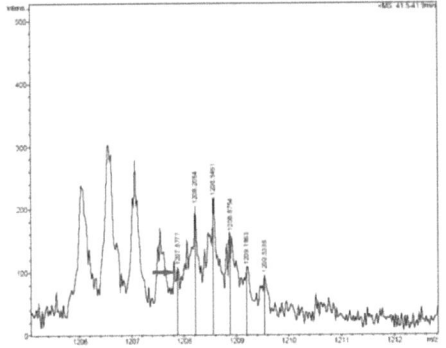 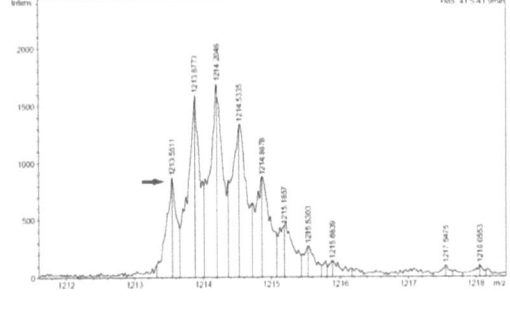

Fig. 46 Mass spectra containing peptide M154 and mass spectra containing peptide M166. Both peptides unoxidized and 1 x oxidized.

Appendix

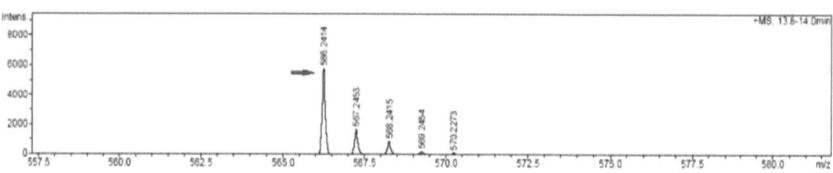

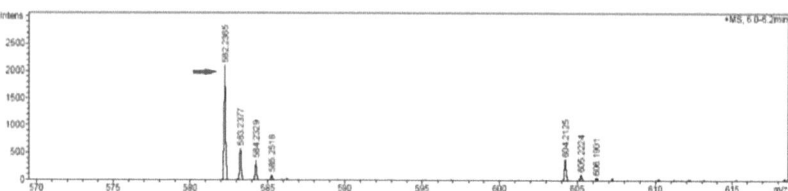

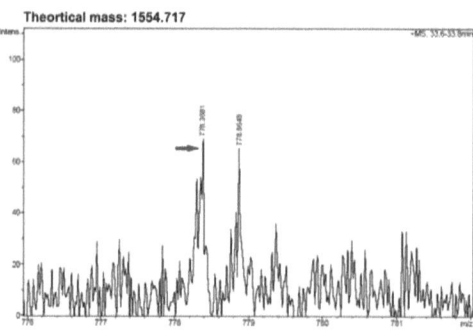

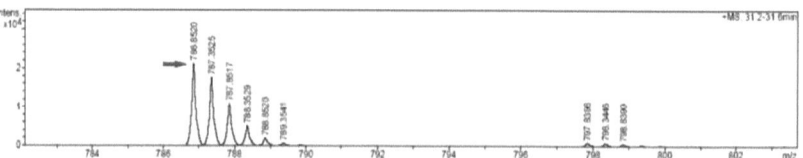

Fig. 47 Mass spectra containing peptide M205/M206 and mass spectra containing peptide M213. Both peptides unoxidized and 1 x oxidized.

Appendix

13.3 Mapping NaIO₄ induced Met oxidation by mass spectrometry - 25 equivalents NaIO₄- pellet fraction

Sequence: TNMK→ M109 [unoxidized]

Theortical mass: 493.237

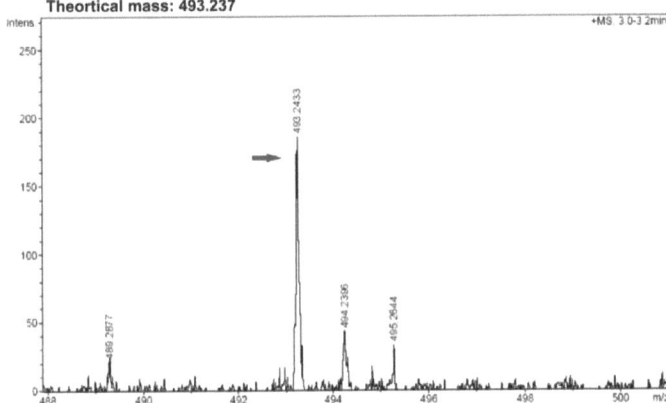

Sequence: TNMK→ M109 [1 x oxidized]

Theortical mass: 509.232

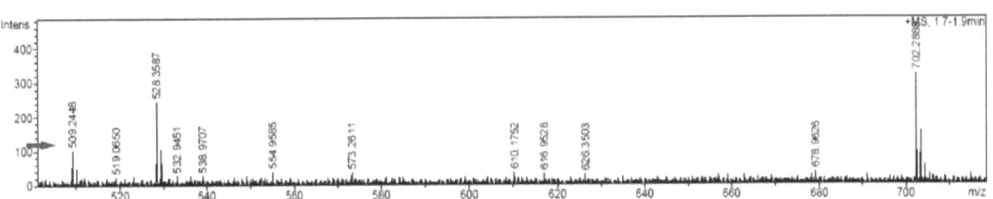

Fig. 48 **Mass spectra containing peptide M109.** Unoxidized and 1 x oxidized.

Appendix

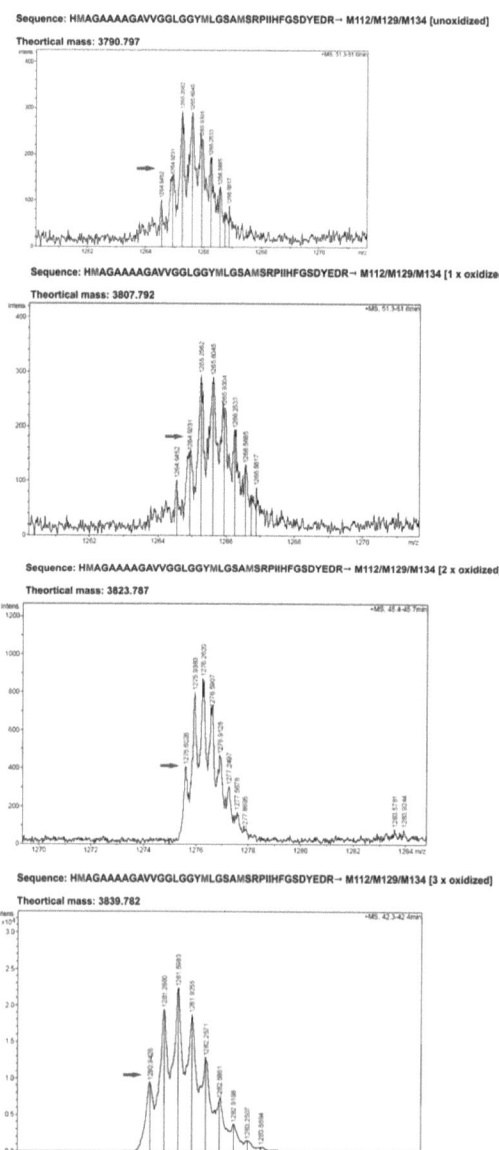

Fig. 49 Mass spectra containing peptide M112/M129/M134. Unoxidized and 1 x, 2 x and 3 x oxidized.

Appendix

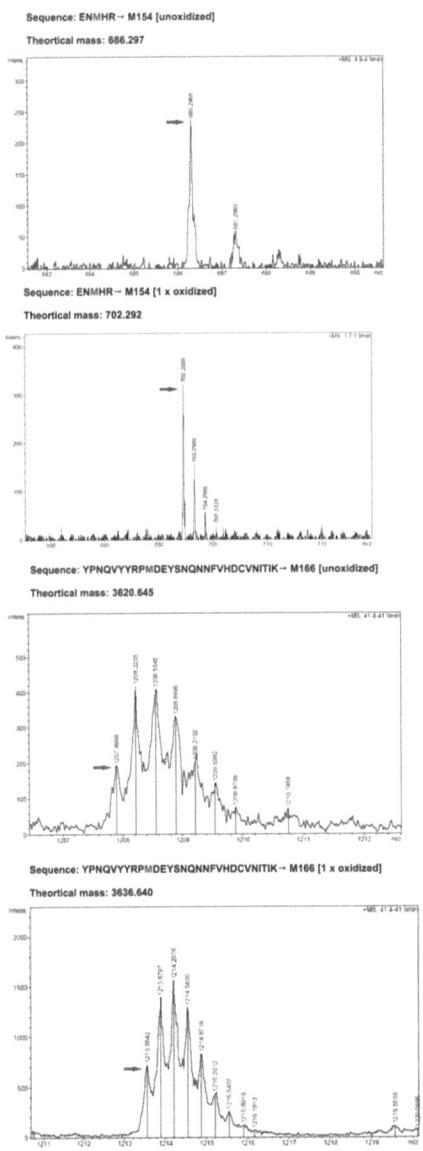

Fig. 50 Mass spectra containing peptide M154 and mass spectra containing peptide M166.
Unoxidized and 1 x oxidized.

Appendix

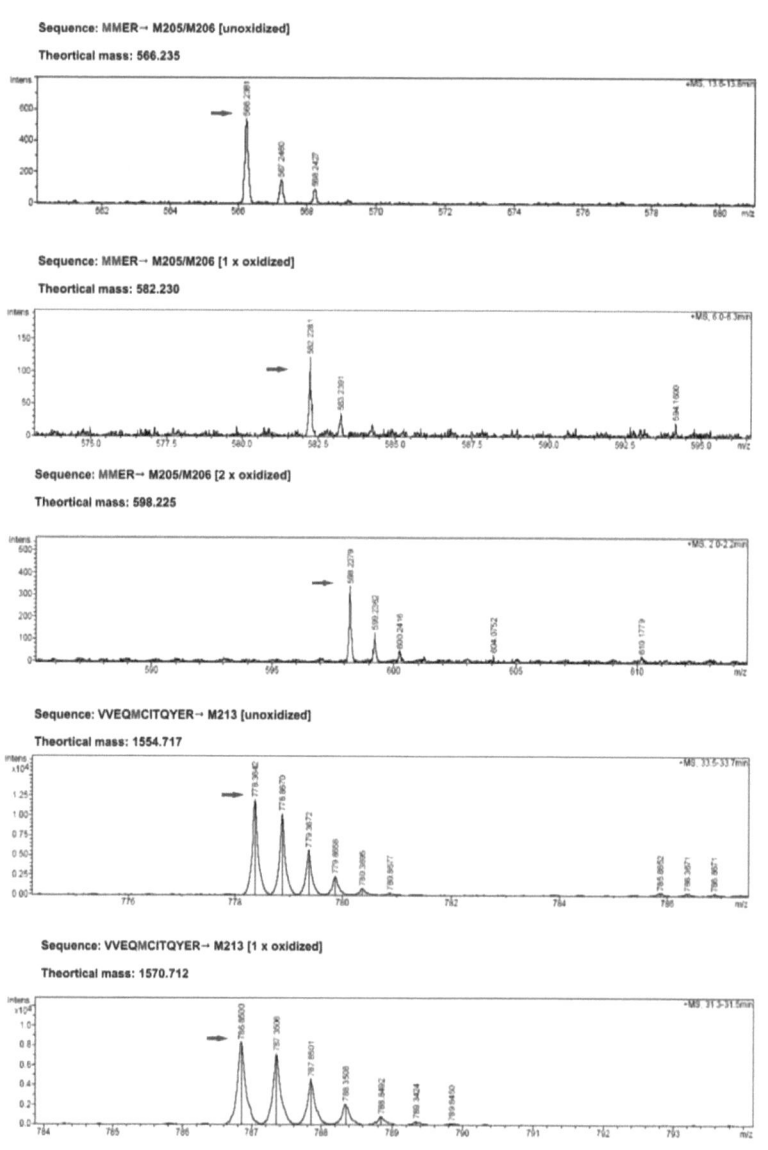

Fig. 51 Mass spectra containing peptide M205/206 and mass spectra containing peptide M213.
Peptide M205 is unoxidized, 1 x and 2 x oxidized. Peptide M213 is unoxidized and 1 x oxidized.

I want morebooks!

Buy your books fast and straightforward online - at one of the world's fastest growing online book stores! Environmentally sound due to Print-on-Demand technologies.

Buy your books online at

www.get-morebooks.com

Kaufen Sie Ihre Bücher schnell und unkompliziert online – auf einer der am schnellsten wachsenden Buchhandelsplattformen weltweit!
Dank Print-On-Demand umwelt- und ressourcenschonend produziert.

Bücher schneller online kaufen

www.morebooks.de

OmniScriptum Marketing DEU GmbH
Heinrich-Böcking-Str. 6-8
D - 66121 Saarbrücken
Telefax: +49 681 93 81 567-9

info@omniscriptum.com
www.omniscriptum.com

Printed by Books on Demand GmbH, Norderstedt / Germany